UFOs and Aerospace

Graphic Illustrations

by

Stephen Watson

Based on UFO concepts of inventor and visionary Gene Watson.

Copyright © 2018 Stephen Watson

All rights reserved.

ISBN-13: 978-1721647019
ISBN-10: 1721647015

The humanity swarm has degree variations (selfish) in every country on earth. The criminal element are chimp mentalities - people looking to take advantage of others for a fast buck. These kinds of minds do not belong in public office.

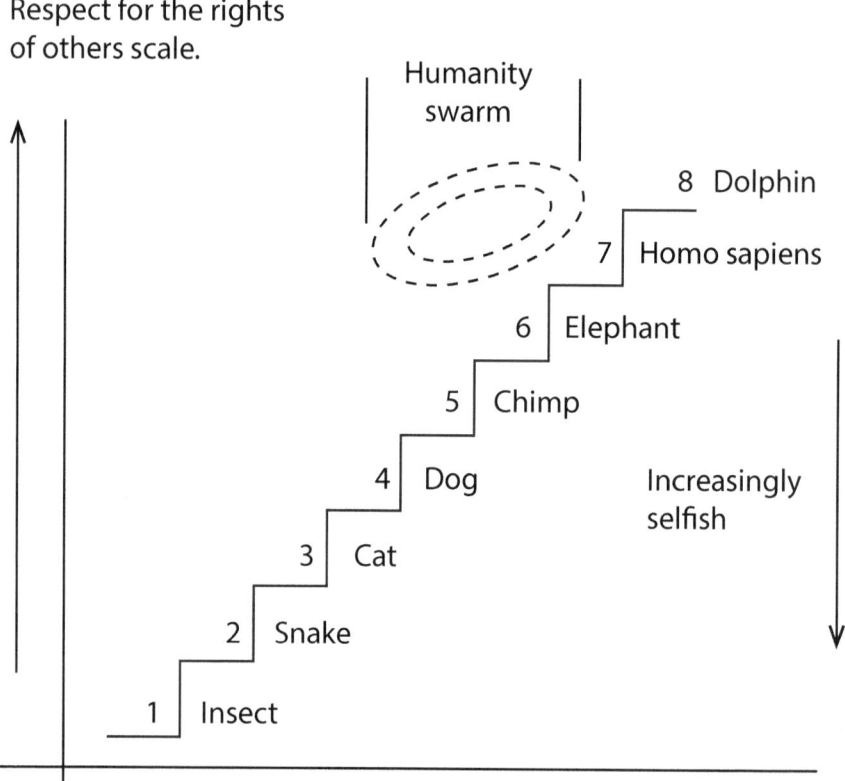

HUMAN PRIMATE PRESS

Table Of Contents - Page 1 of 2

1. The WWII German UFO
2. The German UFO of the 1940s
3. Hamburger-shaped Flying Saucer
4. Flying Saucer - Fission Rocket
5. Flying Saucer - Jet
6. Flying Saucer - Plasma
7. Cylindrical UFO - Electric Field
8. Buoyant UFO
9. AC Ionic Wave Blimp
10. The Wireless Elevator - Beamed Energy Propulsion
11. The Design Dates Back to the 1970s
12. The Egg-Shaped UFO of The 1960s
13. The Egg-Shaped UFO = Fan/Jet
14. German Foo Fighter (1945)
15. Mushroom-Shape UFO - Electric Field
16. Ground Effect Vehicle - Fan, Gear Reduction
17. Top-Shape UFO - Rocket
18. Top-Shape UFO - Electric Field + Rocket
19. The WWII German Atomic Rocket
20. The Triangle UFO - How It Works
21. The Secret Triangle Aircraft - Jet/Rocket
22. Liquid driving a turbine
23. Large Passenger Flying Saucers - How The Fan Can Be Enormous
24. Early Flying Saucers - Jet
25. Piezoelectric Power
26. Piezoelectric Wheel Upgrade
27. Flying Saucer and Jet Function Engine
28. This Eliminates The Solid Stage Boosters For Runway To Orbit
29. Gear Reduction Flying Saucer
30. The 1977 UFO Motor - Ionic, Magnetohydrodynamic
31. UFO Motor - Transparent Ball
32. Can Transparent Ball Inside A Transparent Ball Generate Thrust? Yes
33. What happens if the WC Weirding Field Is Upgraded To Stack? This
34. Transparent Ball Motor Upgrade (Combining Hall Motor With Plasma)
35. UFO Motor - Glass Ball Swivel
36. The Spiral AC Ionic Wave Motor
37. The Spiral AC Ionic Wave Motor Cont.
38. The Spiral Concept Upgrade
39. With Ultraviolet Laser Modification Added: The AC ionic Wave Rocket
40. Permanent Dielectric UFO Power Generator
41. The Possible Nikola Tesla UFO Motor of the 1920s

Table Of Contents - Page 2 of 2

42. To Mars and Back Without Stages
43. Possible Nikola Tesla Motor (Upgrade)
44. Nikola Tesla's Blimp Thruster - 1,000,000 Volts
45. Did Nikola Tesla Invent the Rebalser?
46. Picofarads
47. The WC Strip Discovery
48. The AC ionic Wave Is Corona - Not Lightening Bolt
49. How AC Differs From DC
50. The Flying Saucer Hover Concept
51. What Is Antigravity
52. Understanding Gravity - Inertia and Kinetic Energy
53. Artificial Gravity Flying Saucer For Low Gravity Moons
54. Gravity is a Monopoly Attractive Energy Field (A Created Entity).
55. The Bell Motor As A Electric Power Generator
56. Ice ball Motor - How I Figured
57. Laser Ice Ball Disintegration
58. Artificial Gravity In Space
59. Improved Scramjet Design (Hypersonic)
60. A Fully Retrievable Booster Rocket
61. If the saucer was too advanced (rocket)
62. This motor accounts for some UFO sightings
63. Americium Rocket
64. Counter-Rotating Fan Flying Top Vehicle
65. Counter Rotating Fan Flying Top Vehicle - Few Roads
66. Counter Rotating Fan Flying Top Vehicle - Detail
67. Red Shift

The WWII German UFO

Purpose: Rise vertically and shoot down B-29 bomber fleet.

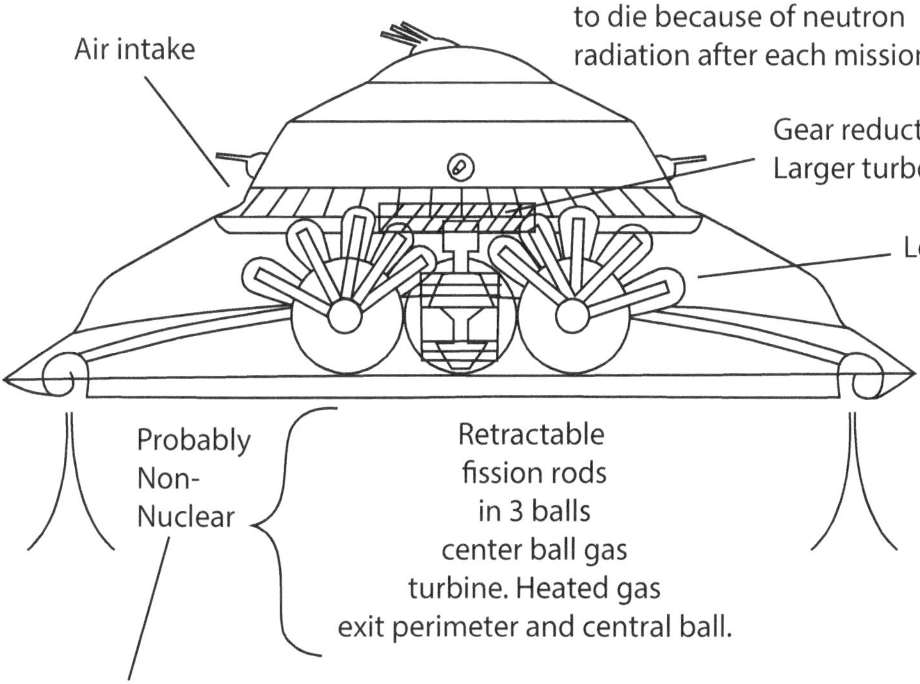

If fission the crew was doomed to die because of neutron radiation after each mission.

Air intake

Gear reduction. Larger turbo fan

Lead

Probably Non-Nuclear

Retractable fission rods in 3 balls center ball gas turbine. Heated gas exit perimeter and central ball.

If so kerosene burn.
Did the Germans build an axial flow fan turbine that powerful (prior to 1945) - To drive a large turbofan (as today's turbojet)?

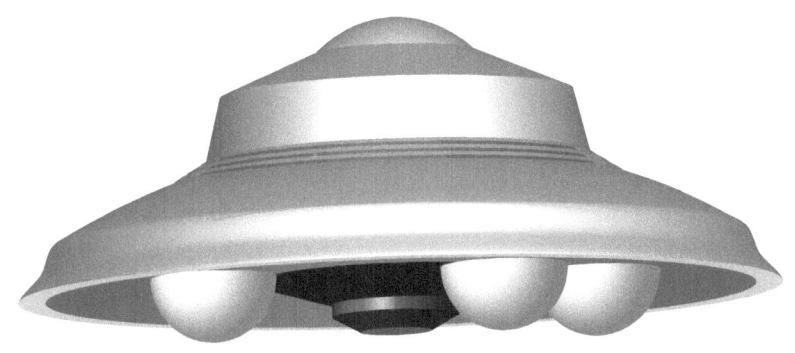

The German UFO of the 1940s

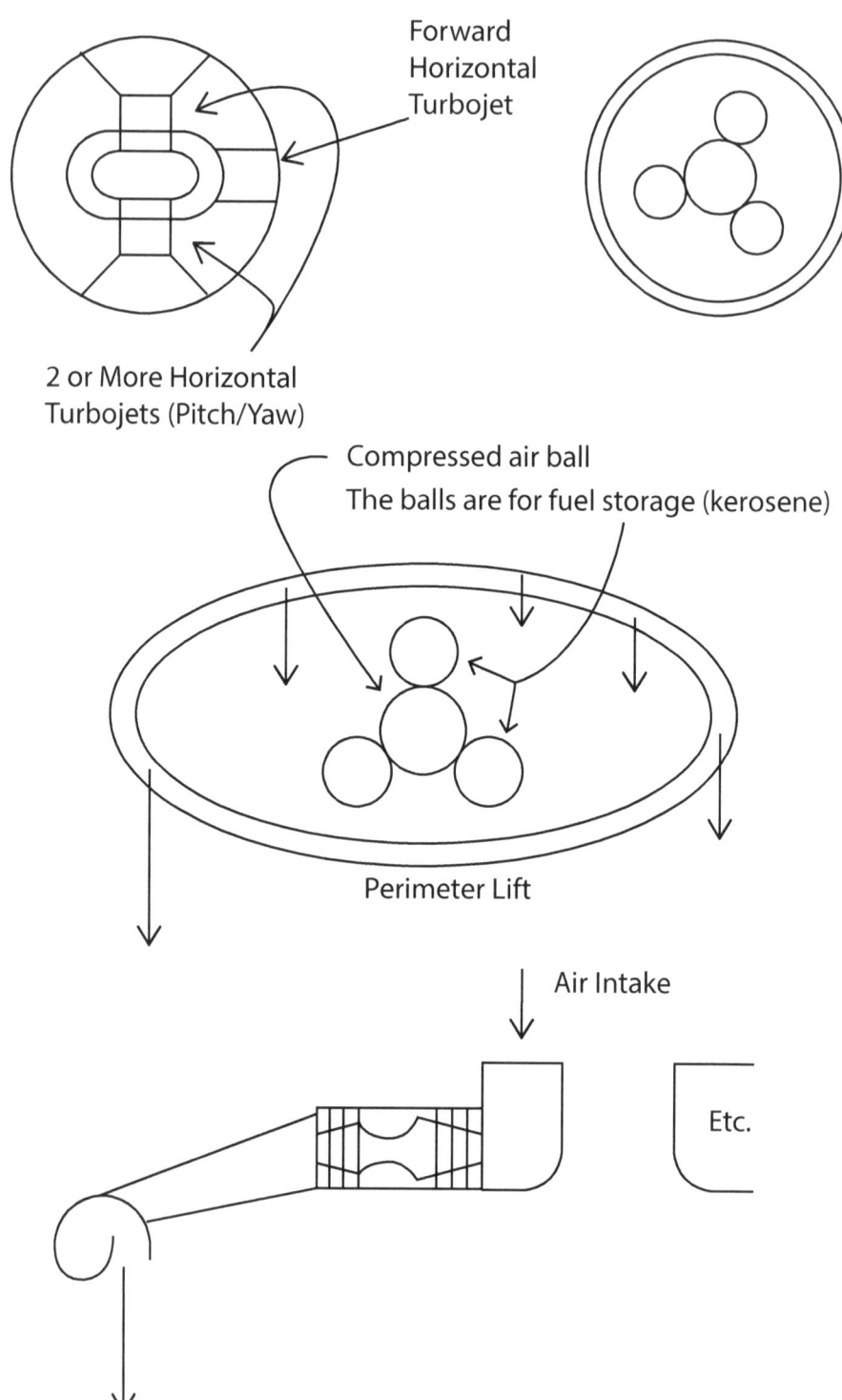

Hamburger-Shaped Flying Saucer

The jet or rocket burn turns into plasma by DC high frequency pulse (arc) which gets accelerated by a superconducting DC magnet.

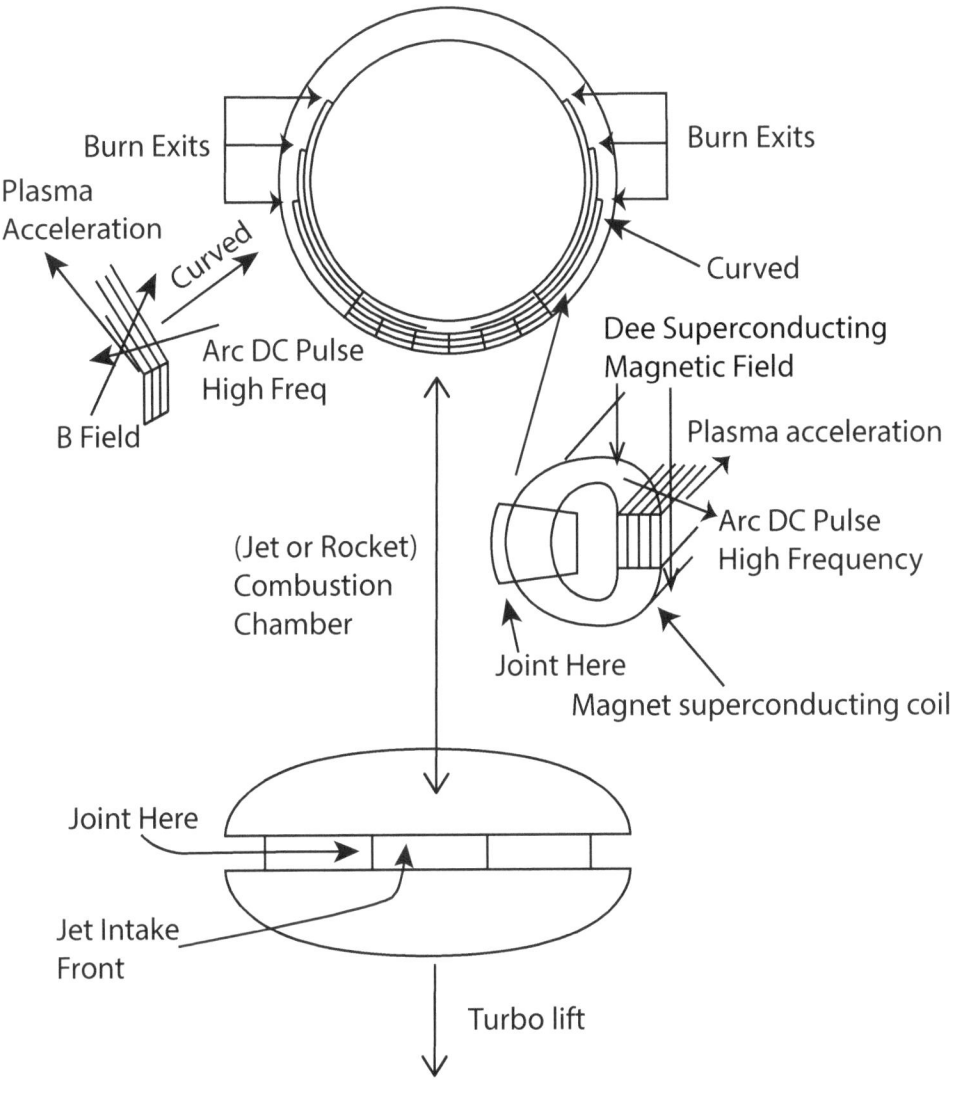

Flying Saucer - Jet/Plasma/Fission Rocket

Put them together and what do you have? Earth liftoff to escape velocity without staging - and can land on Mars.

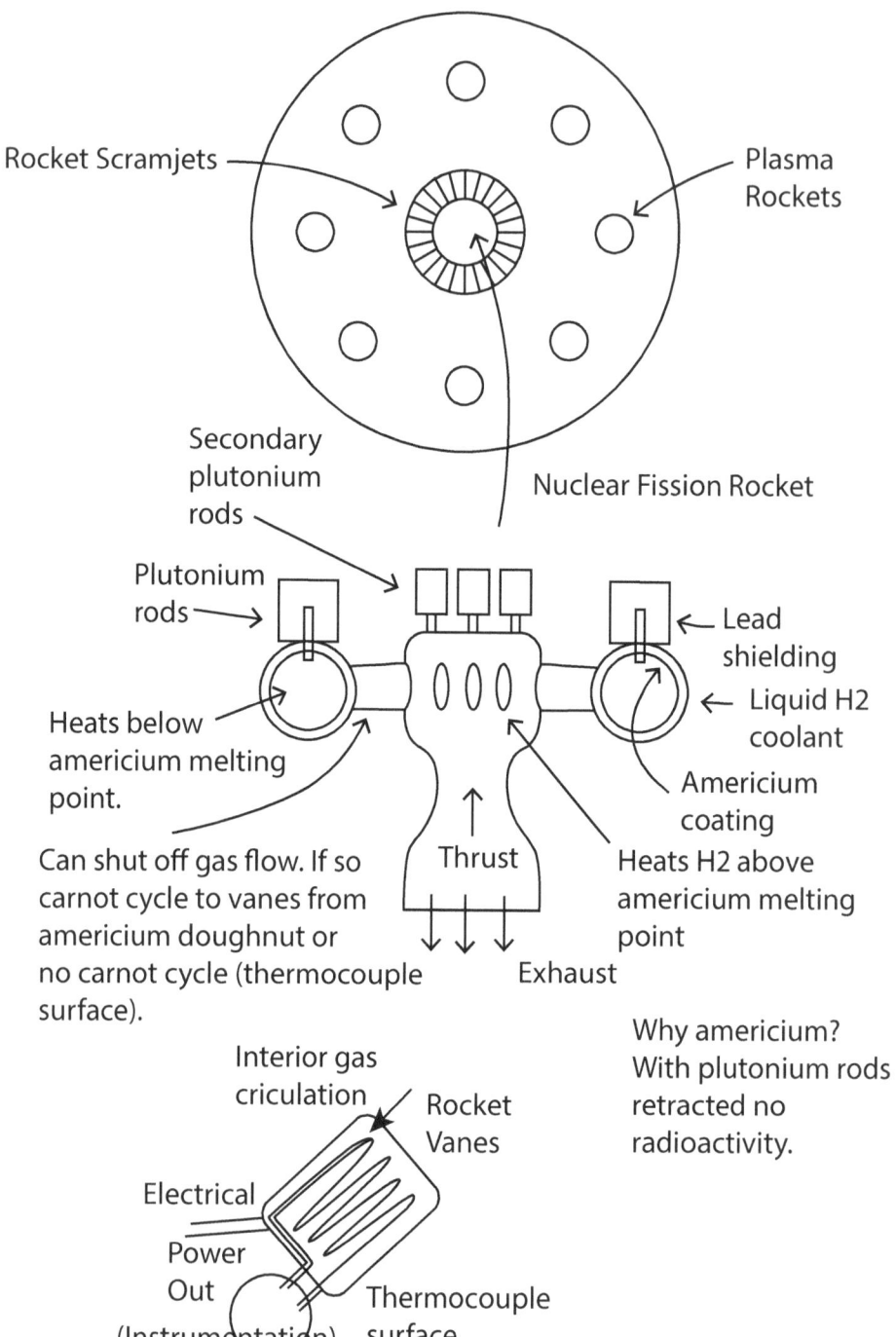

Flying Saucer - <u>Jet</u>/Plasma/Fission Rocket

Put them together and what do you have? Earth liftoff to escape velocity without staging - and can land on Mars.

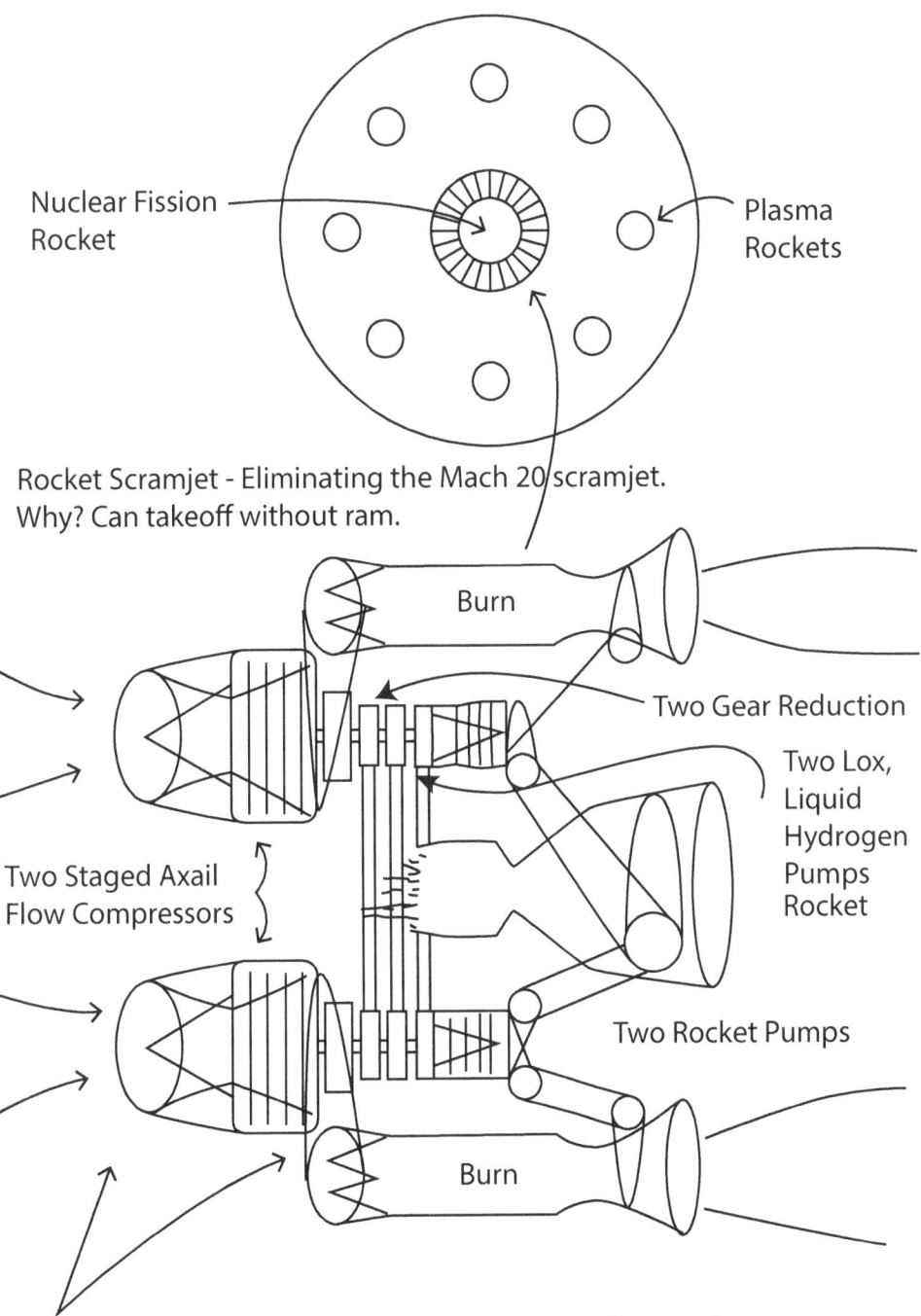

Flying Saucer - Jet/<u>Plasma</u>/Fission Rocket

Put them together and what do you have? Earth liftoff to escape velocity without staging - and can land on Mars.

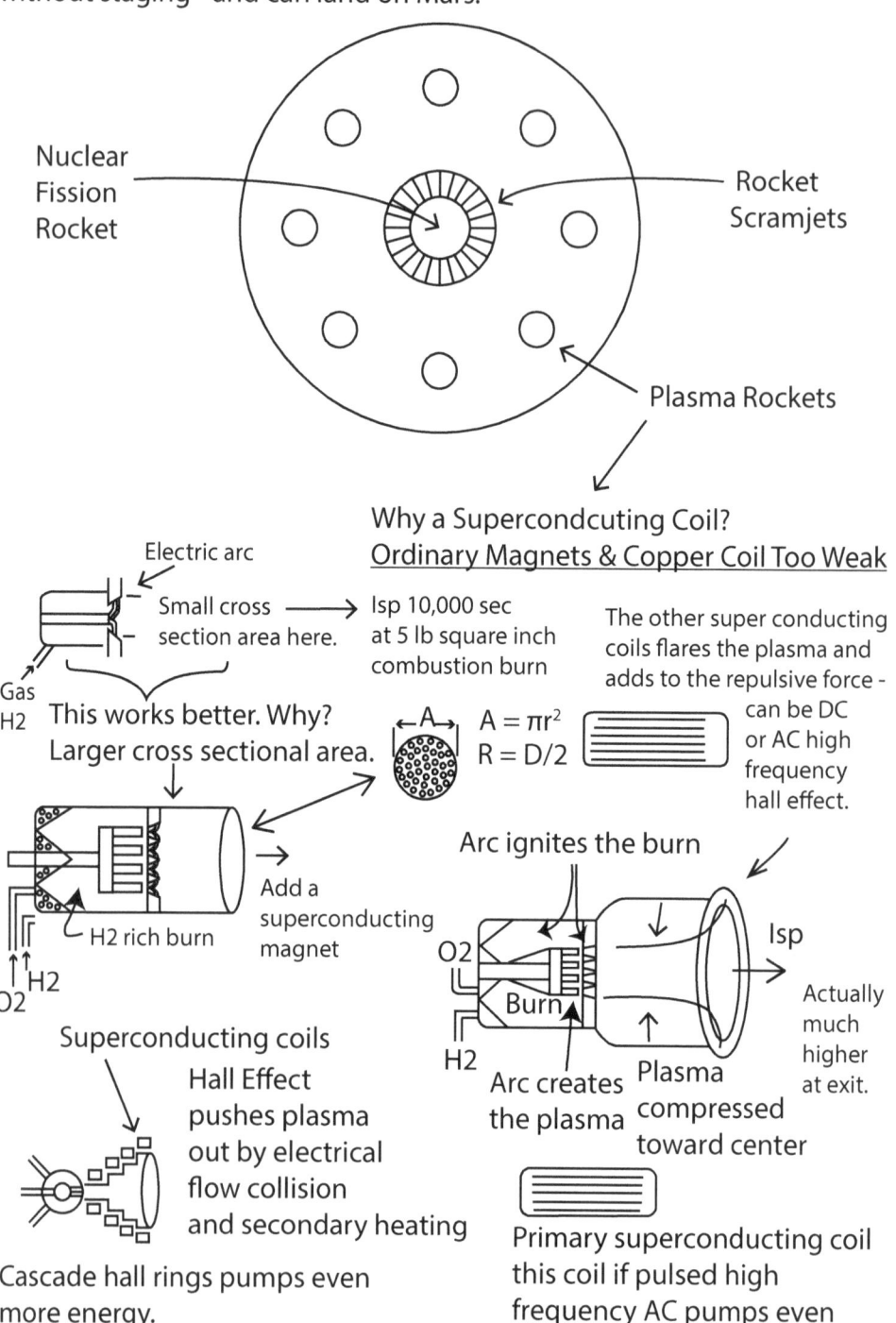

Cylindrical UFO - Electric Field

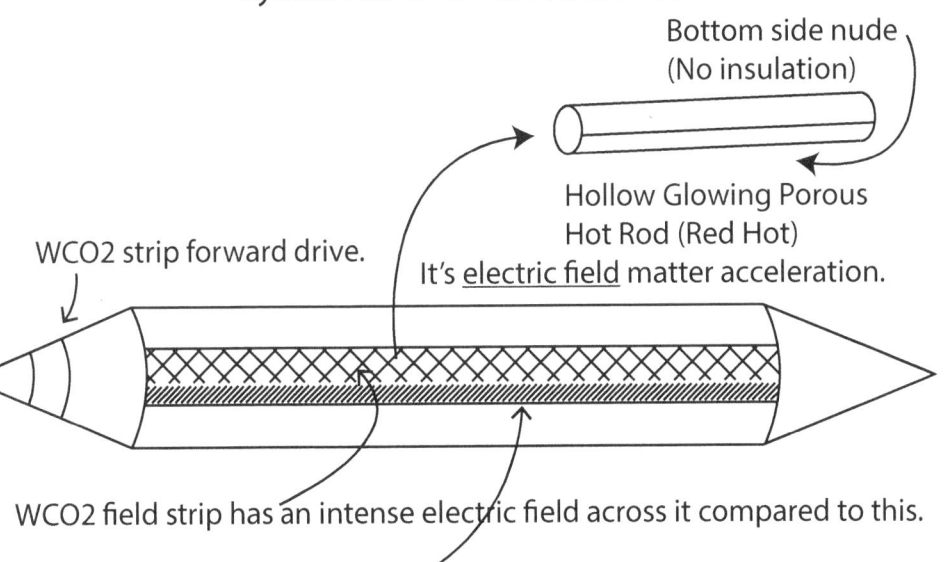

Bottom side nude (No insulation)

Hollow Glowing Porous Hot Rod (Red Hot)
It's <u>electric field</u> matter acceleration.

WCO2 strip forward drive.

WCO2 field strip has an intense electric field across it compared to this.

Rebalser strip 100% insulated coated (no exposed metal anywhere). Will not break down into an arc discharge short at any altitude nude metal strip substitute will not work in AC.

In the atmosphere the gas is compressor pumped into the hollow porous rod. In a vacuum the gas comes from internal storage. It can be liquid.
The formula is $Pe = MV^2$
M = sum of mass accelerated including external atmospheric air charged molecules collide with. If large V = speed drops.
If small (all gas accelerated passing through hollow hot rod) exceeds chemical burn rocket but the applied power.
$Pe = MV^2$ remains constant.

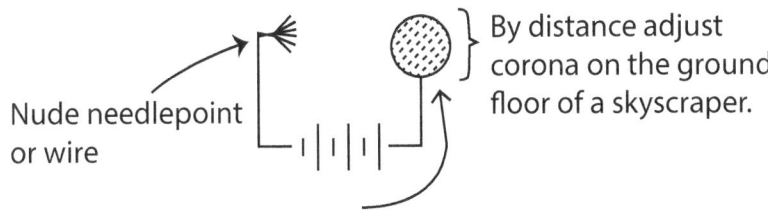

Nude needlepoint or wire

By distance adjust corona on the ground floor of a skyscraper.

Nude Grid the same taken to the skyscraper top floor will break down into <u>arc discharge short</u>.
Rebalser substitute will not work in DC.

Buoyant UFO

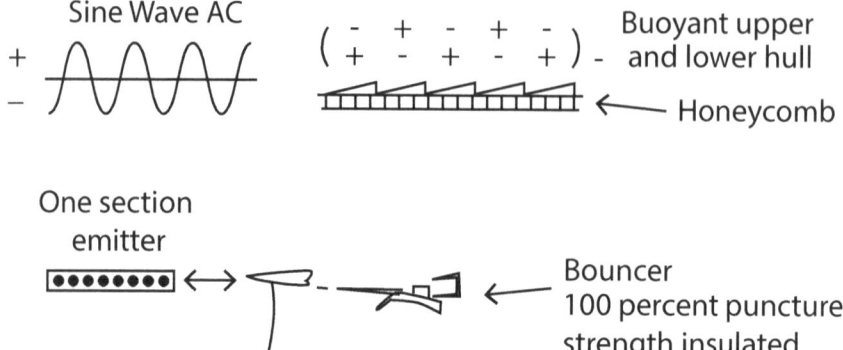

What happens is the ions at the emitter get high voltage electric field acceleration to the opposite polarity at incredible speed but can not arc into the bouncer so it bounces off followed by the opposite polarity wave. This means variable sine wave frequency power supply as the saucer accelerates faster and faster out of the atmosphere into orbit. In space all the emitters use internally supplied gas. If the gas is 20% oxygen and the rest helium the crew has access to the whole interior.

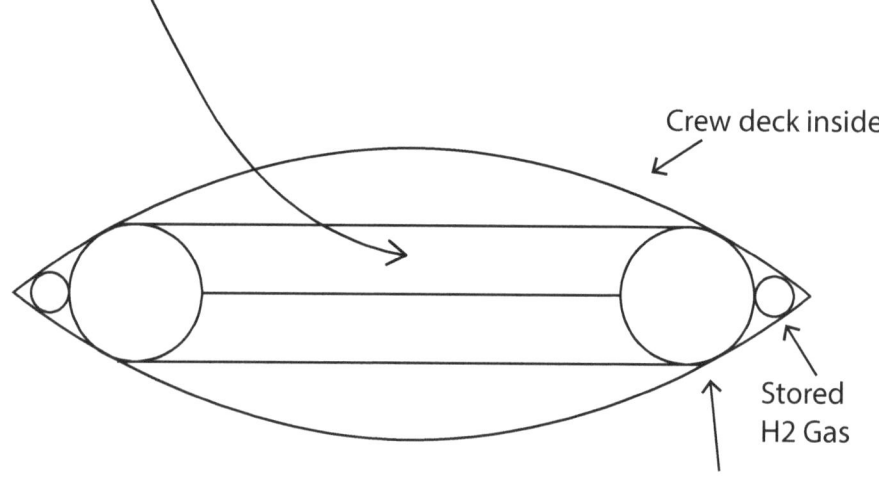

Why buoyant? (Low density) Because the hull has a large surface area. The top and bottom of the hull is theforward drive motor which would be too weak for a high denisty saucer.

AC Ionic Wave Blimp

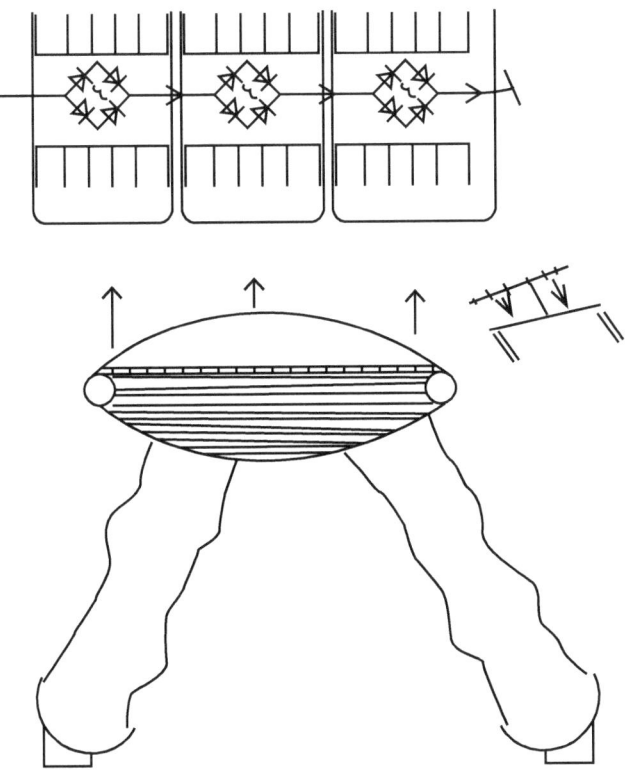

The inside of this can be a lens-shape blimp. A microwave antenna inside of the blimp itself full wedge bridge each of the antenna segment is connected in series you can come up with something like a million volts. That will supply power to the needle points, device and all this will weigh less than the helium that will try and make it lift the hole thing will lift by helium alone and then when you apply microwave power it will go up even higher than the stratosphere and then if you you the kind of conductive ring that lets the gas out of the little holes it will rise all the way out to escape velocity from this planet where the store gas because the lower the atmospheric pressure is the higher the ion ac velocity becomes in a vacuum it far exceeds chemical burn in other words this is capable of ground lift-off to interplanetary travel velocity going straight up with transmitted beams a microwave to it escape velocity beam ship using the AC ionic wave concept.

The Wireless Elevator - Beamed Energy Propulsion

It will lift weight from the ground all the way to 24,500 miles straight up. It drops the cargo in geocentric orbit and comes straight down under its own power for repeated use (flips over for thrust down) then flips up for powered descent.

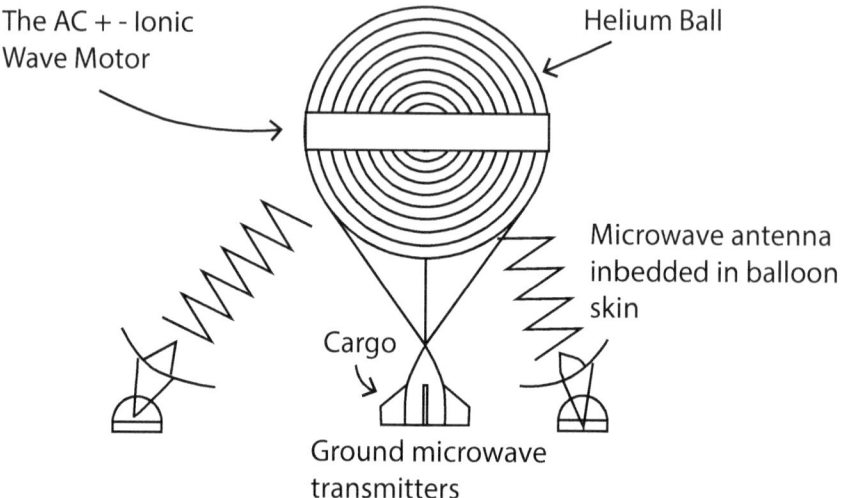

It can be made from already known cargo blimp technology and will not explode in a vacuum.

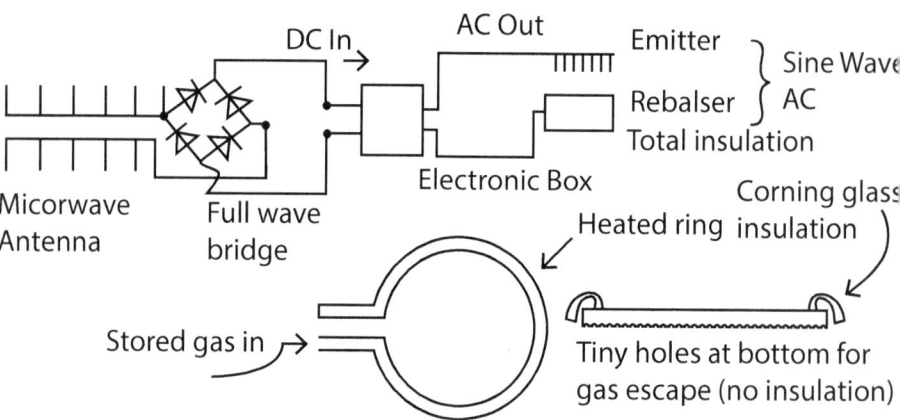

This modification makes it work in a vacuum on stored internal gas as well as in the atmosphere.

This design dates back to the mid 1970s

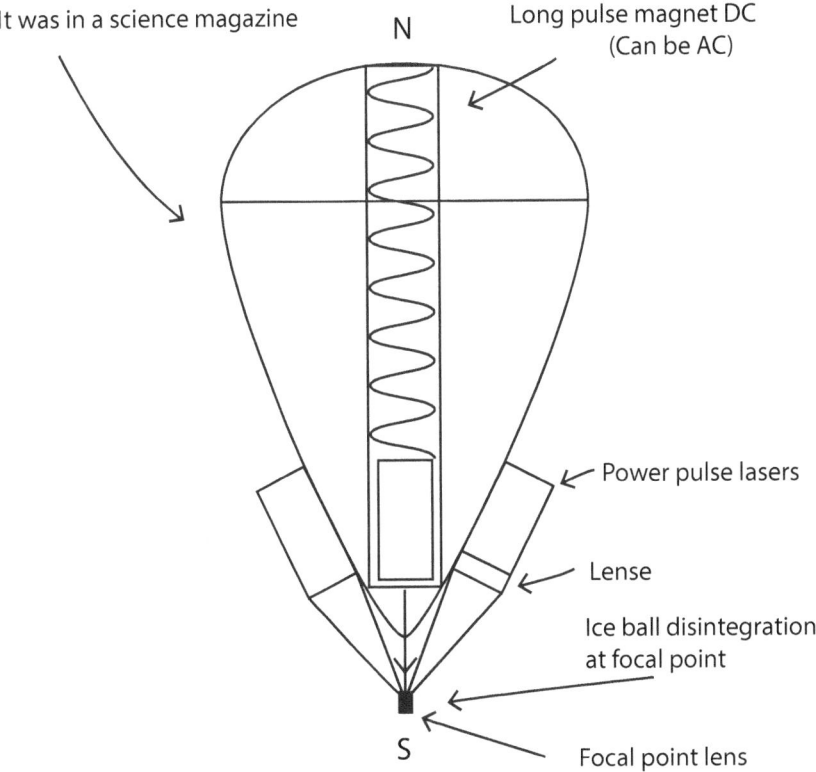

The Egg-Shaped UFO of the 1960s

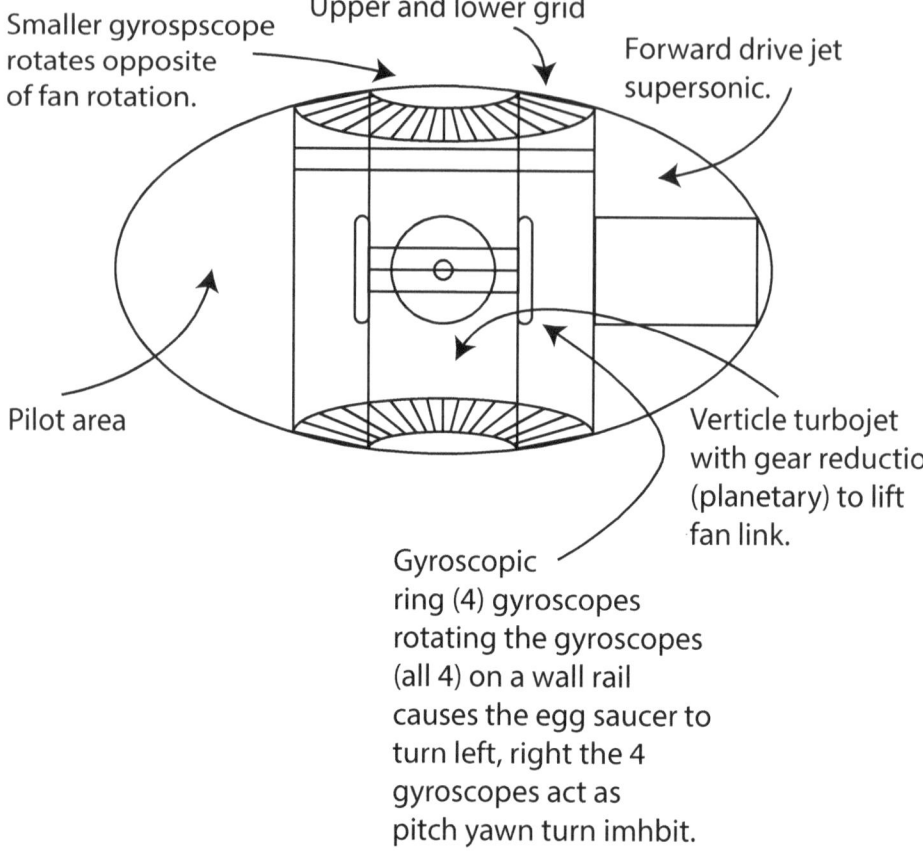

The big breakthrough was the gear reduction link to high speed motors (electric, piston, jet) that resulted in substantial verticle lift power; if electrical it was limited to small models.

The Lunar lander (LM) design in the hands of millionaires gave them a free hand to design a variety of fuselages among which was the egg-shape (elliptical).

The Egg-Shaped UFO - Fan/Jet

Basically a ducted turbofan substitutes for open helicopter blades.

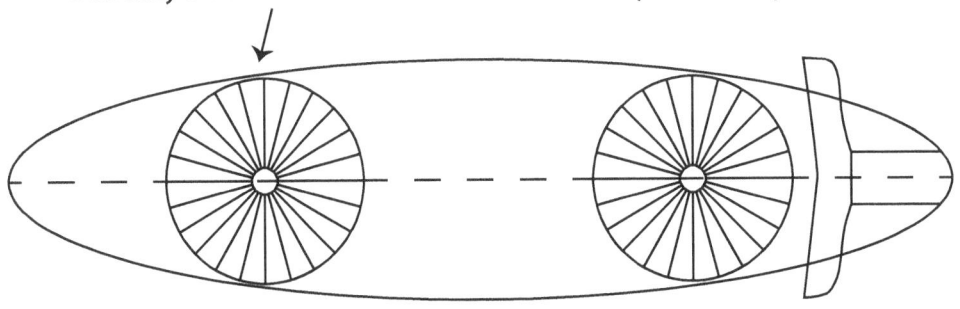

Forward Drive Jet

Pilot Area

Corridor passage permits internal access to front and rear.

Turbine
Gear reduction

Two verticle mounted gas turbines <u>permits corridor passage</u>

Corridor passage

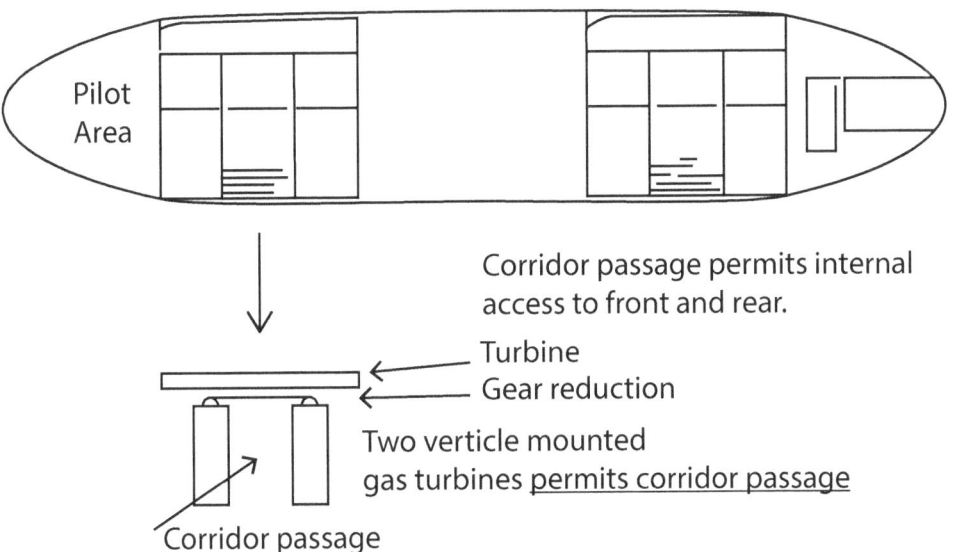

Why a military secret?

{ Advantage - The egg shape UFO is supersonic but can hover like a Chinhook helicopter. It has the Chinhook beat.

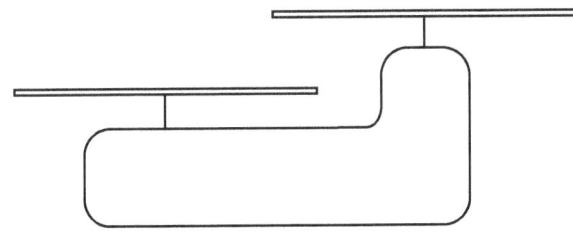

Chinhook helicopter - both use gear reduction gas turbines.

German Foo Fighter (1945)

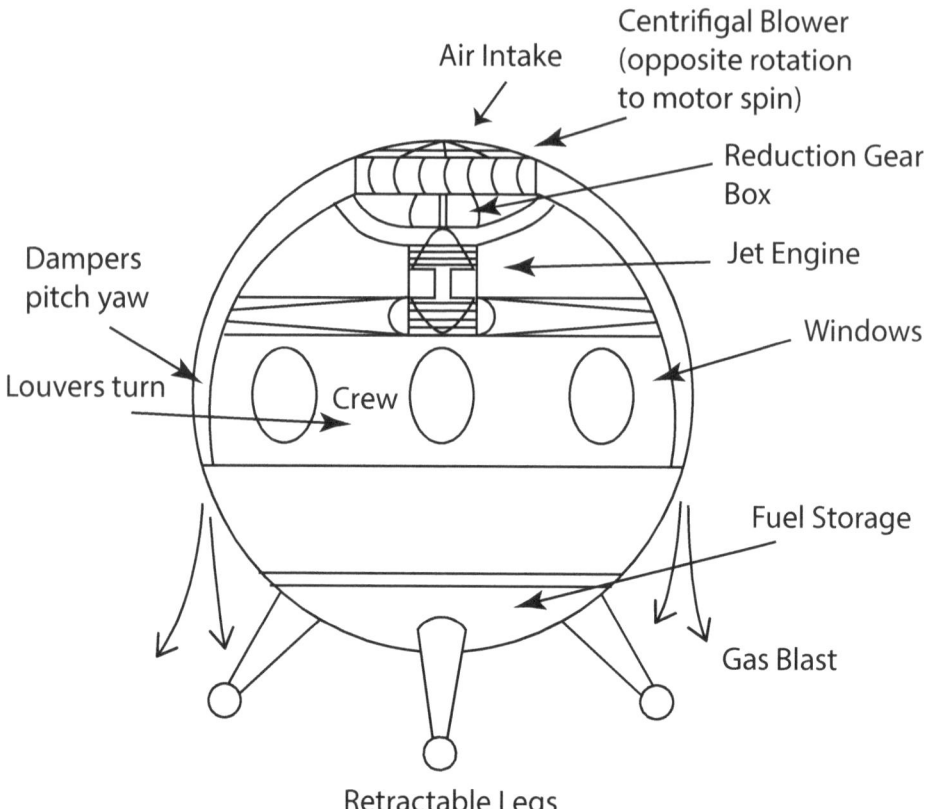

The Foo Fighter was feasible. It would have required a kerosene burn axial flow compressor with gear reduction for vertical lift. Once the trust exceeds weight, wings are not needed. Machine guns would have sufficed to shoot down B-29 fleets. One pilot or two.

Mushroom-Shape UFO - Electric Field

AC Ionic ± Wave Motor
High Pressure = High static thrust
Low Pressure = High exhaust velocity

Theoretically can compete with chemical burn liftoff to orbit

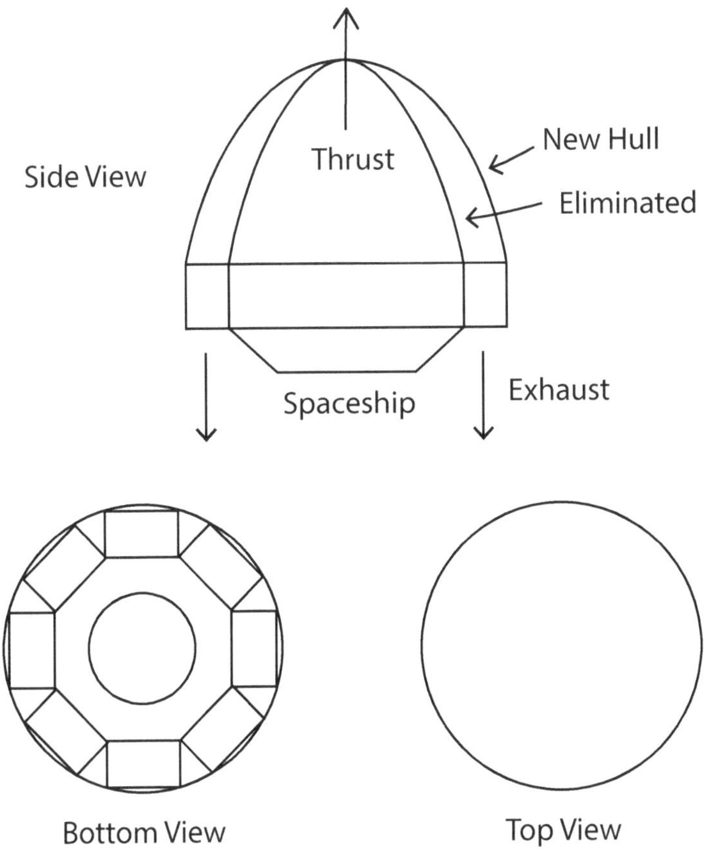

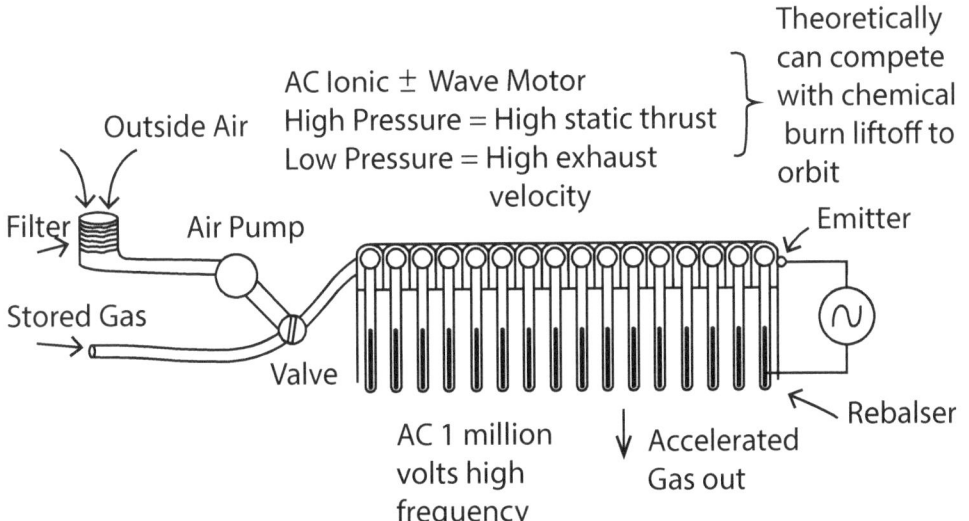

Ground Effect Vehicle - Fan, Gear Reduction

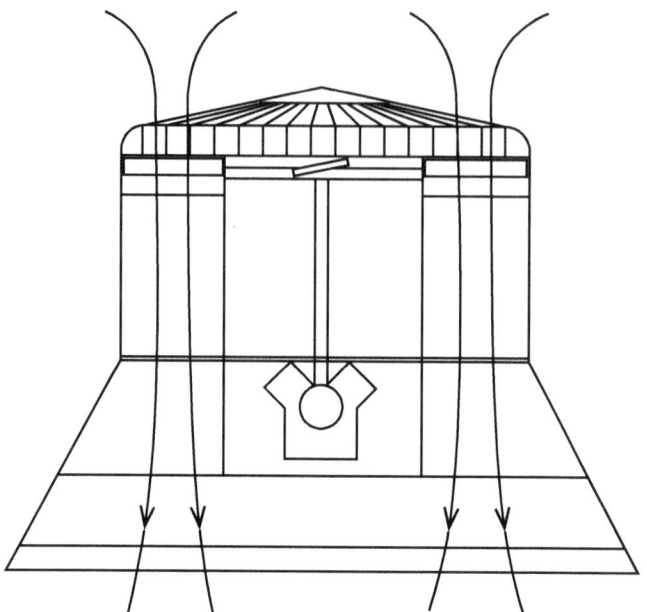

It had a V8 car motor. Electric start connected to a gear reduction shaft. The shaft spun. Why I chose that design is because at low throttle it would dstay in ground effect mode necessary to learn controls and correct glitches.

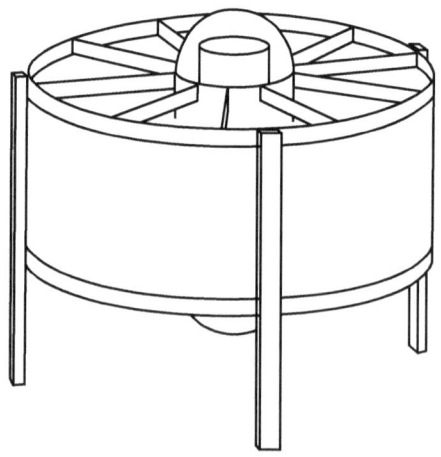

A small model using electric motors and fans.

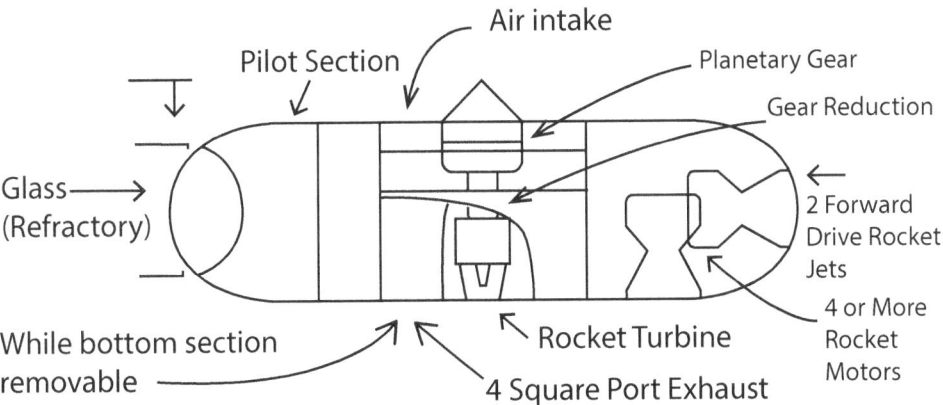

3D laser printing can transform refractory metal oxides; into turbine blades; burn canisters internal combustion piston, rings, walls capable remianing solid excess of 3,000 °c (Iron melts 1,500 °c) The result is an across the board upgrade.

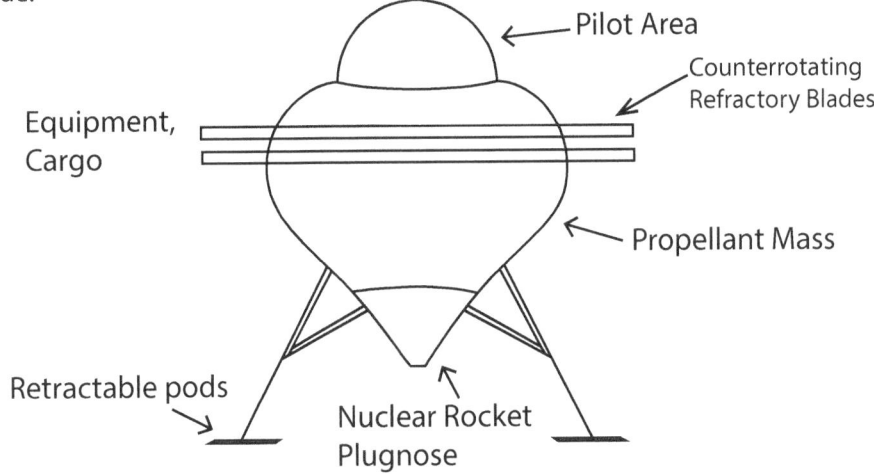

This configuration eliminates parachute re-entry from orbit. The nuclear rocket is drone because it takes this design where propellant mass is a neutron shield to protect the crew until the nuclear rocket shuts off. The top shape uses hydrogen propellant in order to shield astronauts from neutron exposure. The nuclear fission rocket raises the isp from 360 to over 800. Add to that laser 3D [printing] that melts the highest refractory metal oxide powders and the fission rocket can attain an isp of about 2000 with no significant loss of static thrust. The refractory lift blades can lift the UFO nuclear top into the stratosphere before firing the nuclear rocket. Plus soft land on Mars. On Mars with nitrogen extracted from the atmosphere it can go more than orbital. It can go well past escape velocity because mars gravity is one third G. The first test would be a variety of jet and chemical burn prototypes.

Top-Shape UFO - Electric Field + Rocket

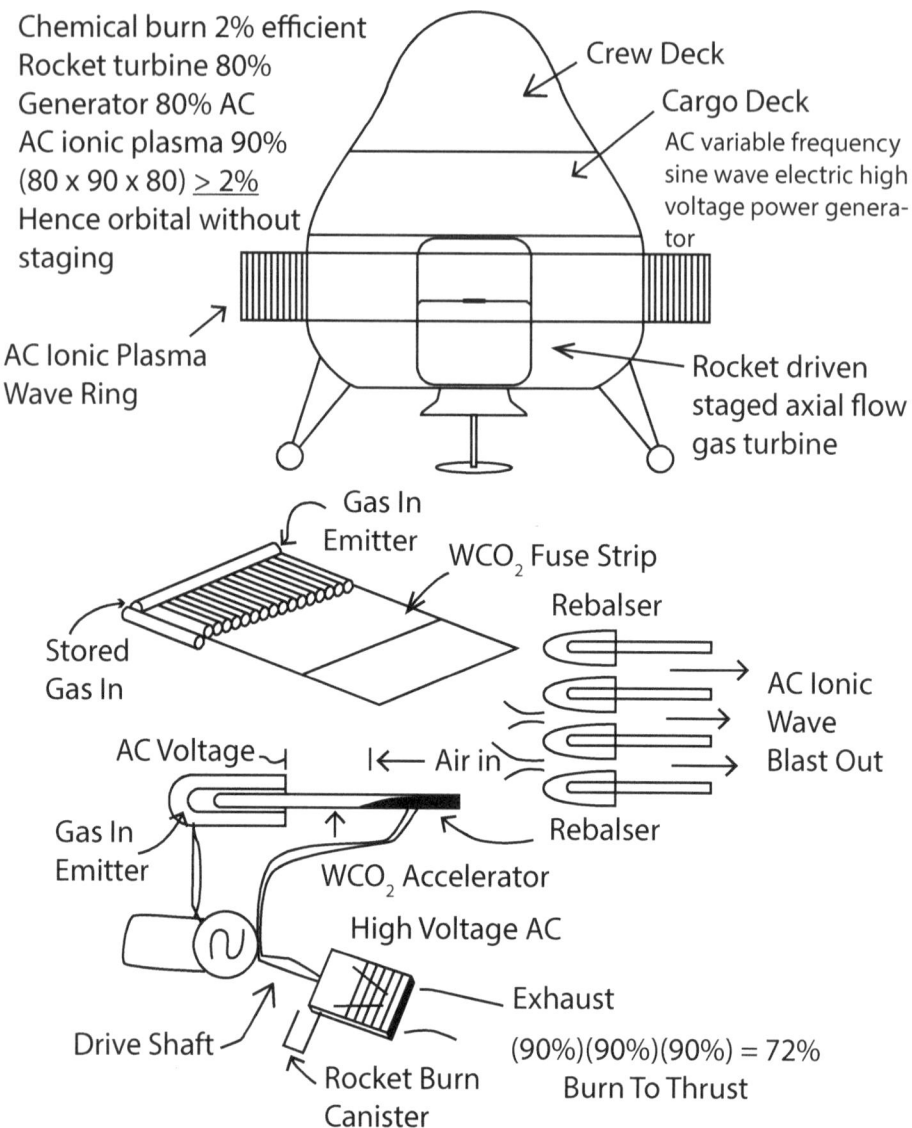

Chemical burn 2% efficient
Rocket turbine 80%
Generator 80% AC
AC ionic plasma 90%
(80 x 90 x 80) ≥ 2%
Hence orbital without staging

Crew Deck
Cargo Deck
AC variable frequency sine wave electric high voltage power generator

AC Ionic Plasma Wave Ring

Rocket driven staged axial flow gas turbine

Gas In Emitter
WCO$_2$ Fuse Strip
Rebalser
Stored Gas In
AC Ionic Wave Blast Out
AC Voltage
← Air in
Gas In Emitter
Rebalser
WCO$_2$ Accelerator
High Voltage AC
Drive Shaft
Exhaust
Rocket Burn Canister
(90%)(90%)(90%) = 72% Burn To Thrust

This device can be curved around the shuttle craft with vertical liftoff to orbit. The rebalser can't be punctured hence there is no arc breakdown. Increasing applied voltage increases static thrust and past 60 miles the internal gas valves to the AC emitter kicks in for uninterrupted acceleration to orbit. Where does that extra energy come from? Well, the chemical burn rocket is only two to 3 percent efficient. The turbine driver is 60 to 80 percent efficient. The super conducting AC sine wave generators are up to 90 percent efficient and the AC wave motor efficiency is also up to 90 percent. With all that combined it beats the 2 percent efficiency of a chemical burn rocket, so it can go from earth to orbit without staging.

The WWII German Atomic Rocket

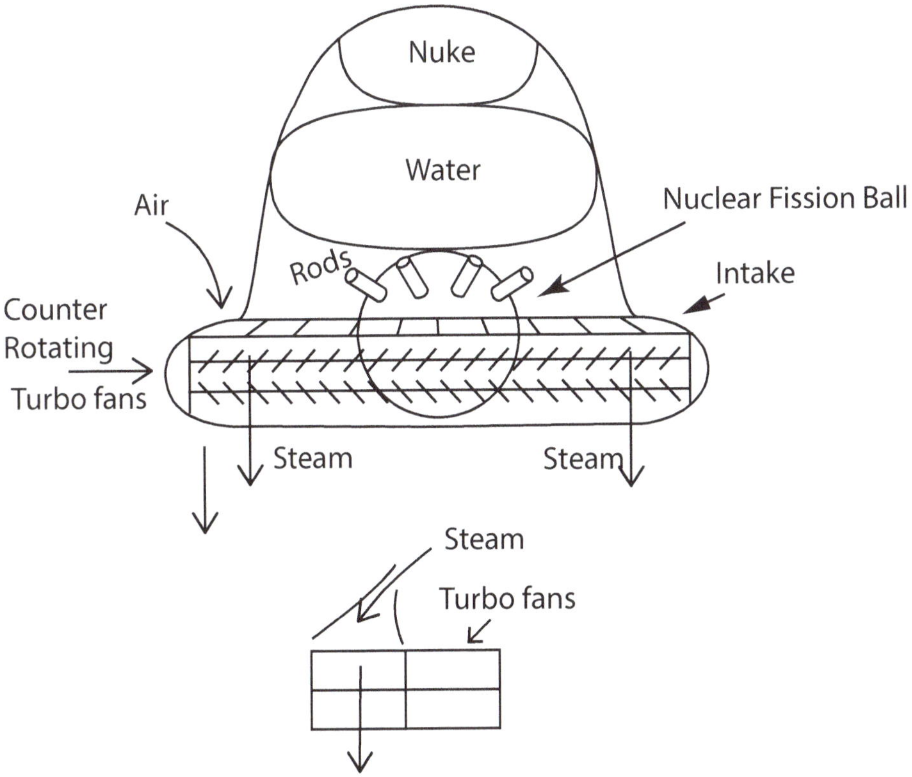

Diagram labels: Nuke, Water, Air, Rods, Nuclear Fission Ball, Intake, Counter Rotating Turbo fans, Steam, Steam, Steam, Turbo fans

The German UFO had water for propellant mass. Water was pumped into the fission ball - transforming into steam. The steam drove the trurbo fans. The vehicle rose like a V2 rocket. The war ended before the Germans could accumulate sufficient fission uranium to create plutonium in quantity. What was discovered was is their launch pad, drawings but no completed device.

The Triangle UFO - How It Works

Combination Verticle Lift Craft { Nuclear Jet / Nuclear Rocket } Fission

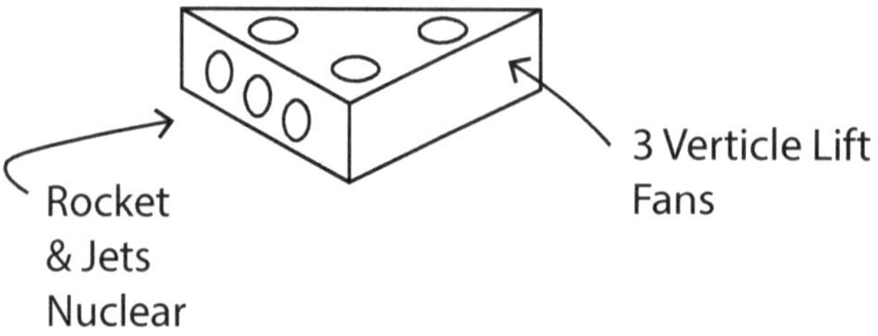

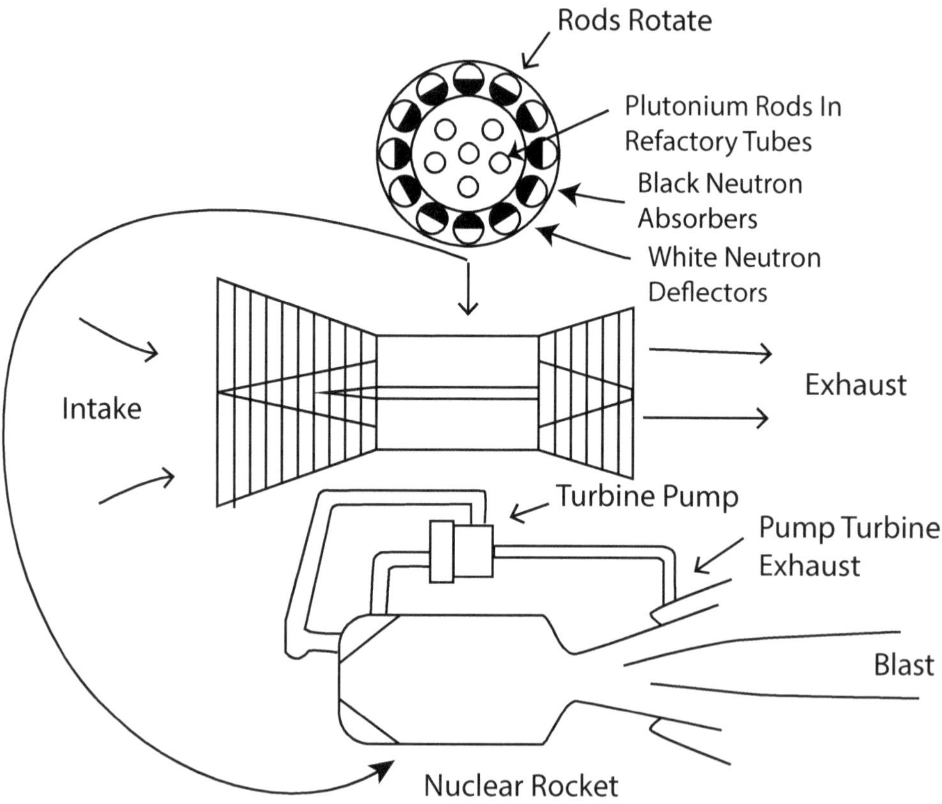

The Secret Triangle Aircraft - Jet/Rocket

Objective: With a series of upgrades increasing maximum speed to orbital. Beyong that nuclear power.

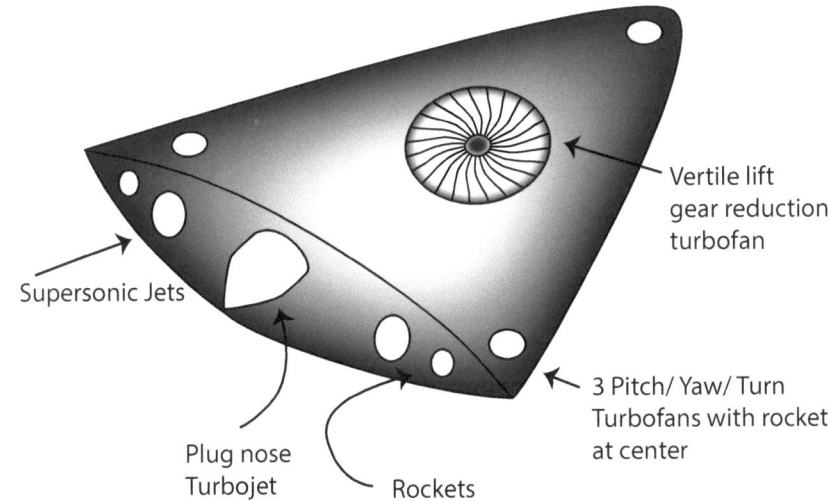

Supersonic Jets

Vertile lift gear reduction turbofan

3 Pitch/ Yaw/ Turn Turbofans with rocket at center

Plug nose Turbojet

Rockets

A big center thruster plus three turbofans at the three angles. The there is a rocket jet forward thrust potentially orbital.

My experiment in 1968 Samuels ave. Ft Worth, Texas.

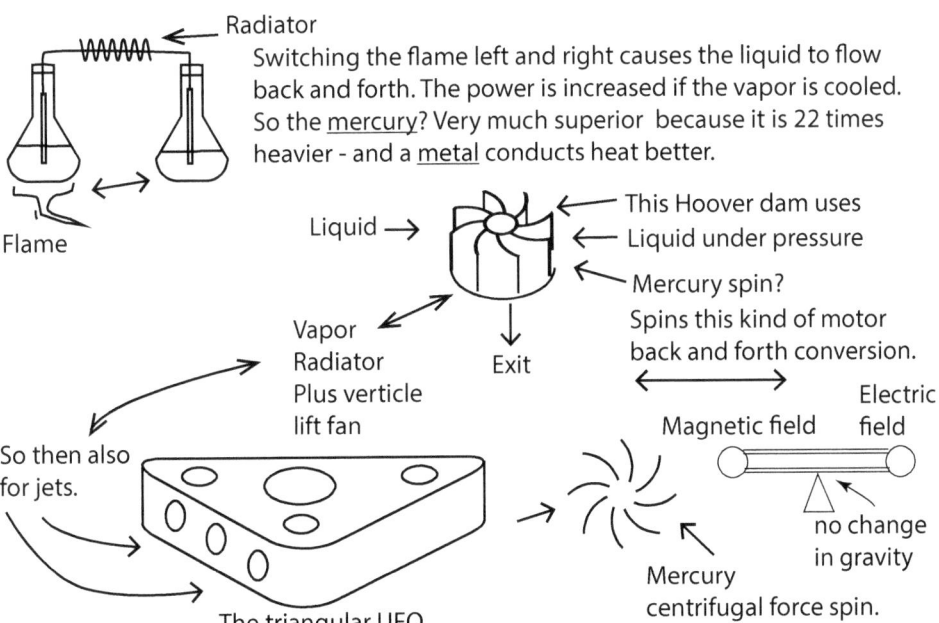

Radiator

Switching the flame left and right causes the liquid to flow back and forth. The power is increased if the vapor is cooled. So the <u>mercury</u>? Very much superior because it is 22 times heavier - and a <u>metal</u> conducts heat better.

Flame

Liquid →

This Hoover dam uses Liquid under pressure

Mercury spin? Spins this kind of motor back and forth conversion.

Vapor Radiator Plus verticle lift fan

Exit

Magnetic field   Electric field

So then also for jets.

The triangular UFO

Mercury centrifugal force spin.

no change in gravity

87% weight loss -> Nuclear energy heating mercury as a liquid reduces weight for power out makes the whole thing weightless.

The spiral liquid under pressure turbine Hoover DAM DESIGN works so well that no one has improved it.

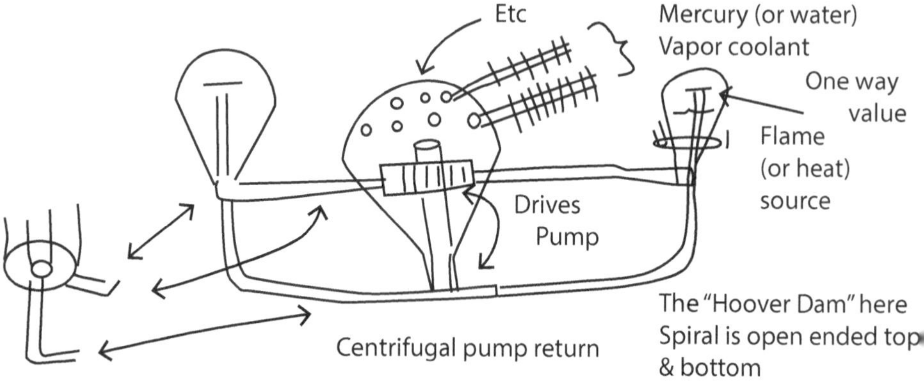

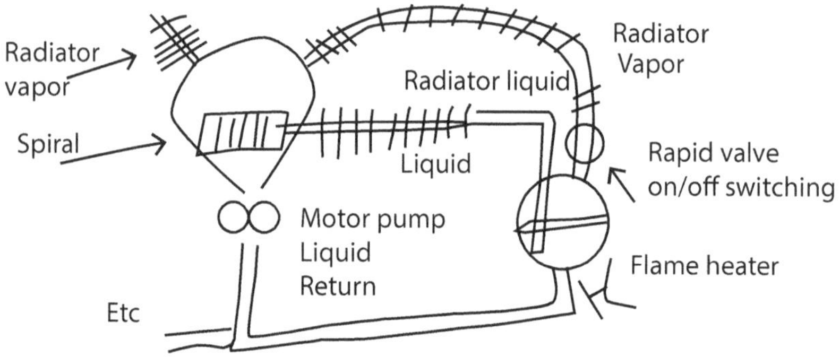

Basically - The old design (steam locomotive) used vapor to drive a piston by boiling wtaer. (1968) I demonstrated liquid driving a turbine is more efficient but the secret is the actual detail. I gave the idea away to an engineer. Carnot cycle mercury is the working fluid the torque is 22 times Higher than water under pressure. So then Mercury liquid driving a turbine vastly reduces size and weight. Hence it makes sense to valve Mercury vapour gas pressure rapidly instead of the old system of using a gas to drive a turbine. I got the hacker monkeys off my back.

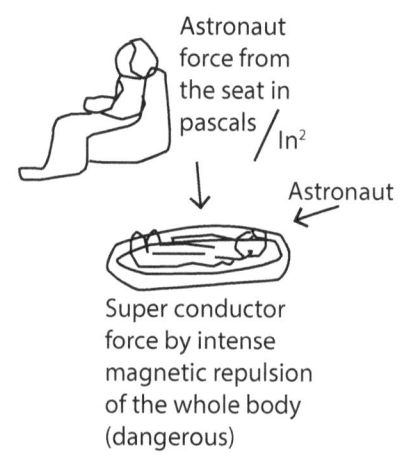

Large Passenger Flying Saucers
How The Fan Can Be Enormous

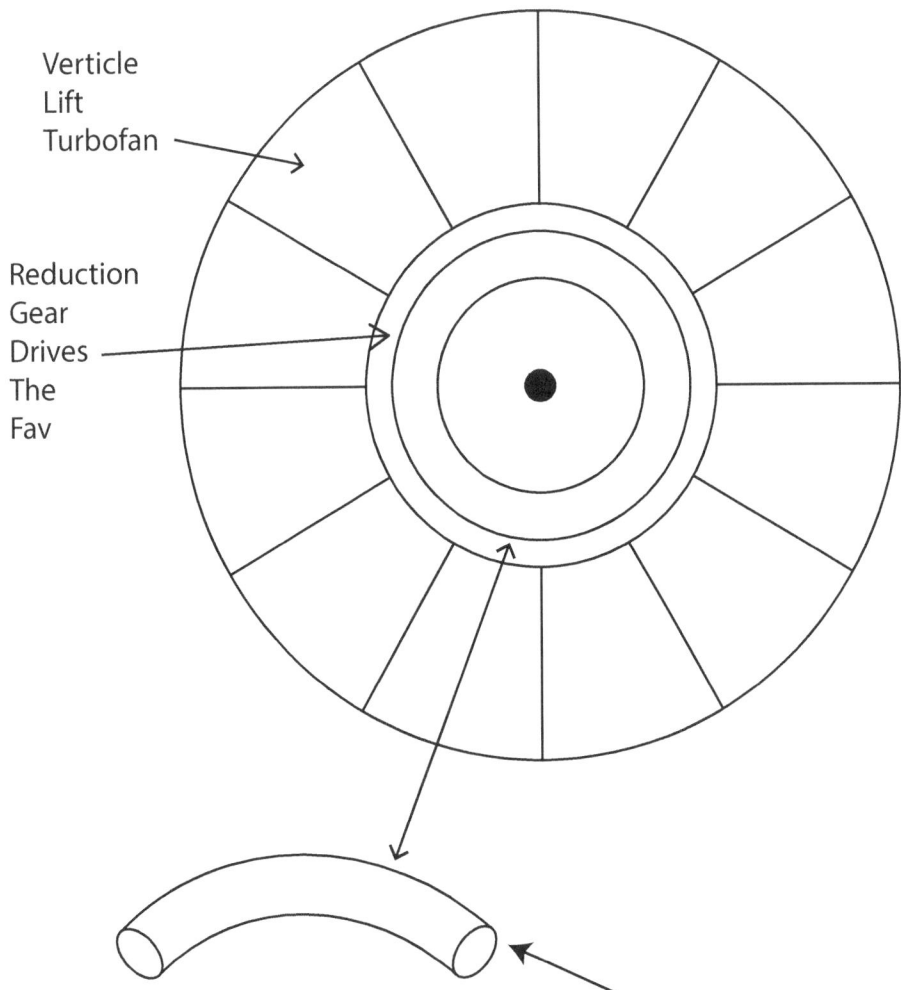

A few ball bearings inside the circular tube

The ball bearings arrange themselves by centrifugal force to cancel vibrations making the fan self balancing. The big vertical lift fan [pictured] is balanced by conventional means to which is added the circular tube. This means the vertical lift fan can be enormously big. If so the saucer can be bigger than the biggest passenger planes in use today today.

Early Flying Saucers - Jet

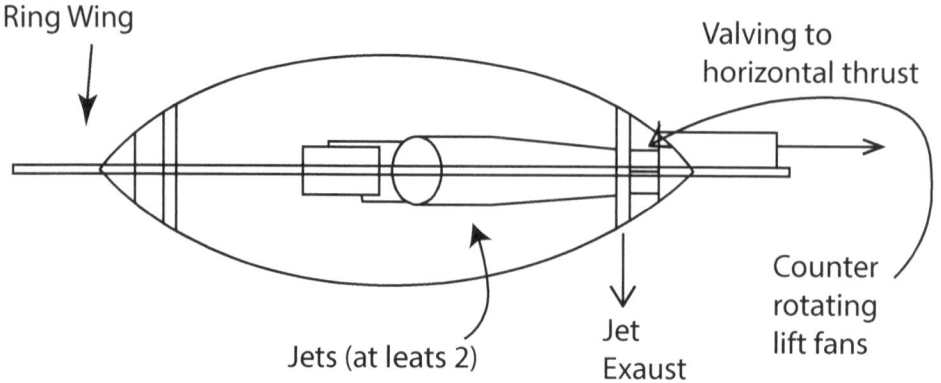

The Nazi innovation was to substitute the Italian [Campini] Caproni centrifugal blower (1939) with the staged axial flow compressor. This improved the performance of the jet engine by raising pressure in the burn canister but the hot gasses driving the rear turbine gave the jet at the time problems. It is because steel softens and becomes like taffy near its melting point - They did not have the metal alloy that made the Harrier Jet possible - So the Nazi's opted to use the jet exhaust to drive the lift turbine (counterrotating) in the configuration of a flying saucer. By valving the jets it diminished vertical lift but increased forward thrust. There were a lot of crashes. Why? Because a big lift fan must be perfectly balanced.

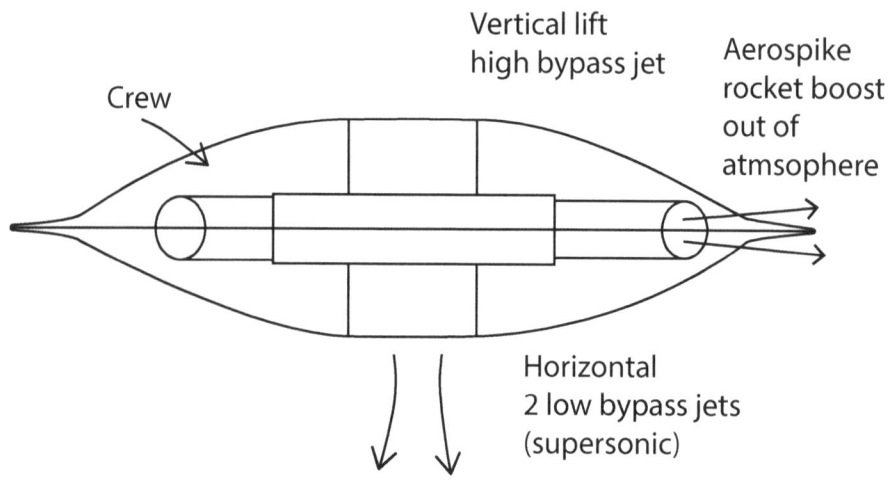

Piezoelectric Power
Simple 1 Piezoelectric Disc Metal Plates Both Sides.

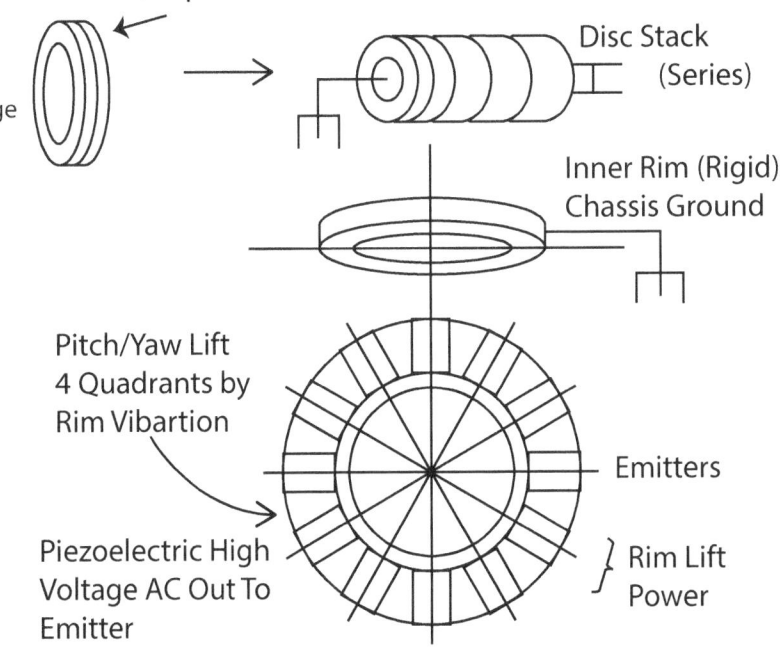

Lift Rim Subsitutes For Solid Stage Booster Rockets

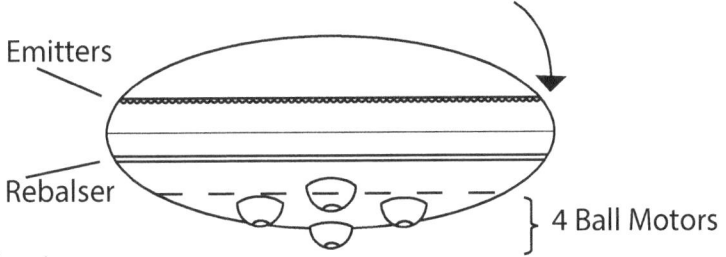

Rim Lift Accelerates
External air down creating lift. The four ball motors are specialized internal gas AC ionic wave (corona) thrusters that continue to push the saucer vertically to interplanetary travel speeds.

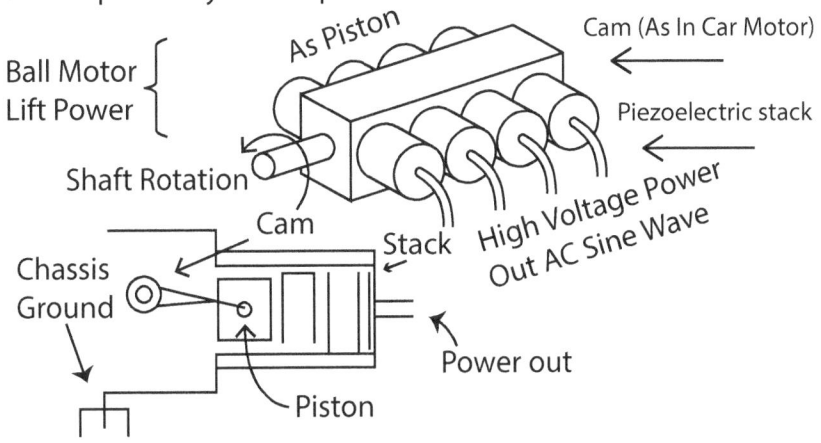

Piezoelectric Wheel Upgrade

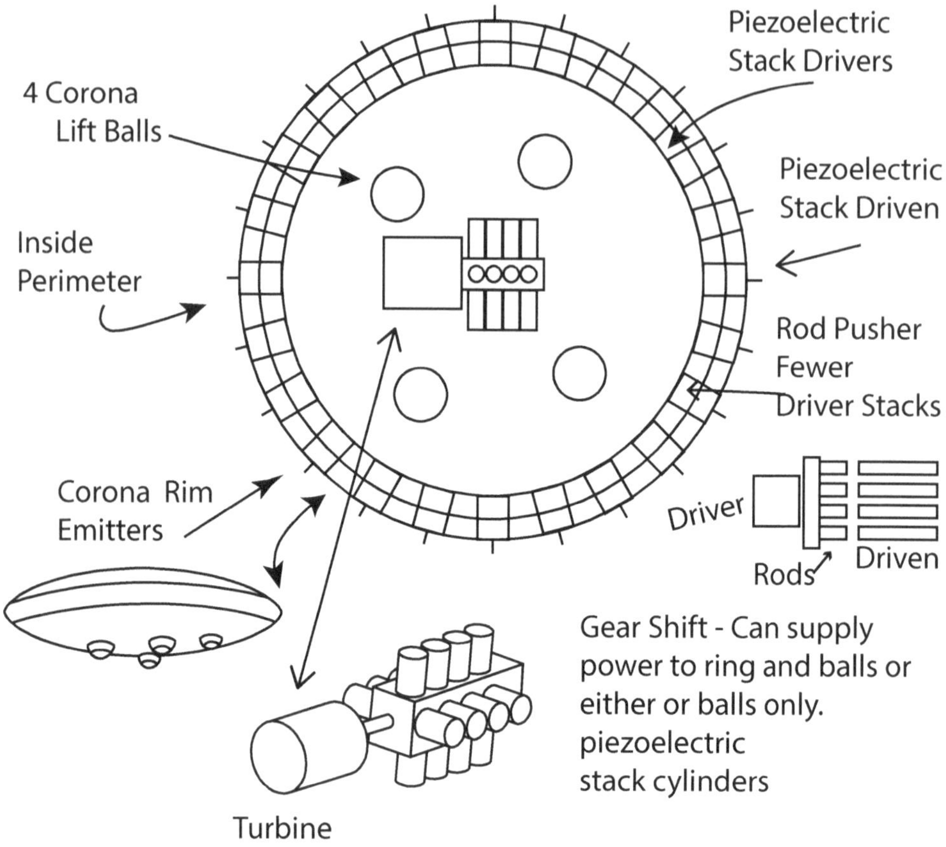

- 4 Corona Lift Balls
- Inside Perimeter
- Corona Rim Emitters
- Piezoelectric Stack Drivers
- Piezoelectric Stack Driven
- Rod Pusher Fewer Driver Stacks
- Driver Rods / Driven
- Gear Shift - Can supply power to ring and balls or either or balls only. piezoelectric stack cylinders
- Turbine

This eliminates the vibration bands. The rigid ring is tripple - the inner piezoelectric stack gets its power from the inner turbine driver cylinder stack because the piezoelectric stack is also a driver by applying AC current which vibrates the driven stack.

So What? Transforming

- Ring 2 actually flat
- For parallel shaft passage
- Ring 3
- Driver shorter fatter piezostack
- Ring 2
- Longer stacks driven
- Ring 1
- Higher AC voltage out
- Holes for shafts ring 2
- Shaft vibrates through holes

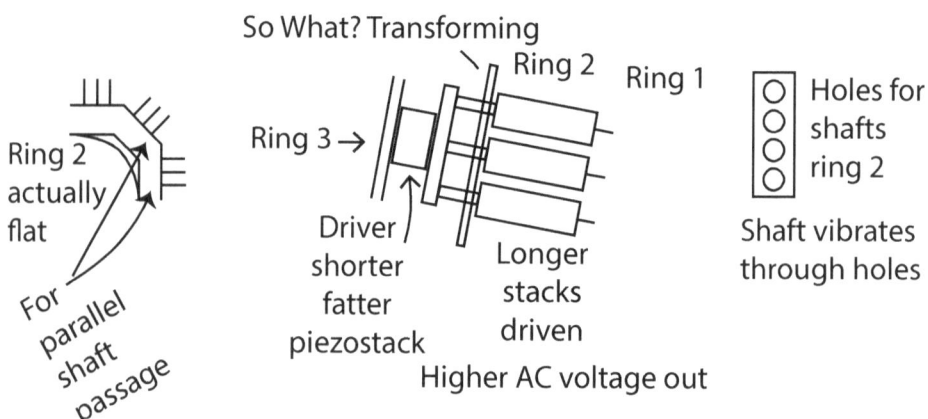

Flying Saucer - Rocket and Jet Function Engine

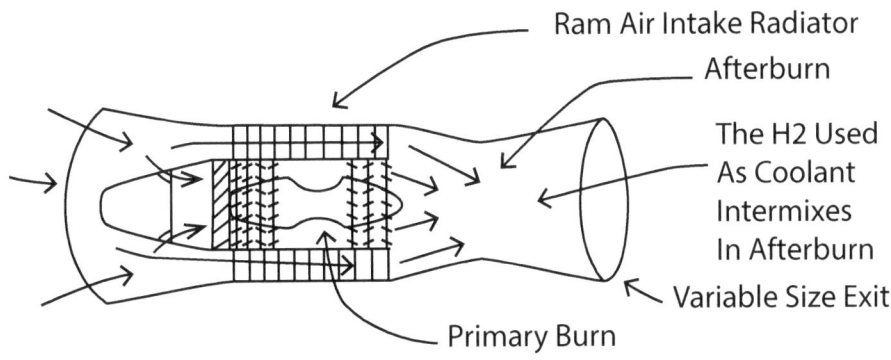

This intake is designed to close off past 60 miles altitude where the motor continues to function like a rocket. Where <u>Liquid hydrogen</u> cools the ram air which afterburns at the axil flow turbine exhaust. Computer calculations demonstrate runway takeoff to <u>orbital velocity</u> from this engine alone (Mach 18). Fuel consumption is only a smidgen of what a chemical burn rocket consumes. To develop the engine to full capacity it requires interfacing the rocket and jet function by degrees, plus a fuselage that can withstand high mach speeds.

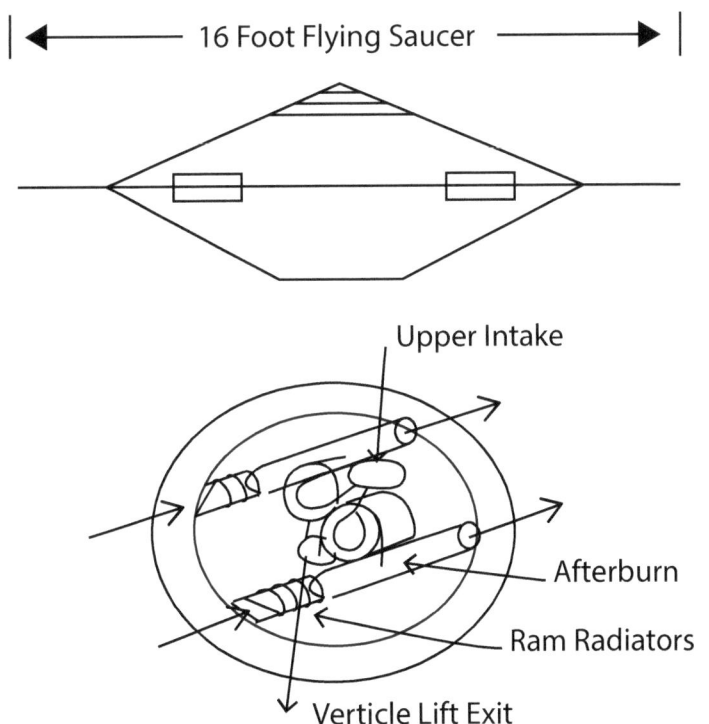

This Eliminates The Solid Stage Boosters For Runaway To Orbit

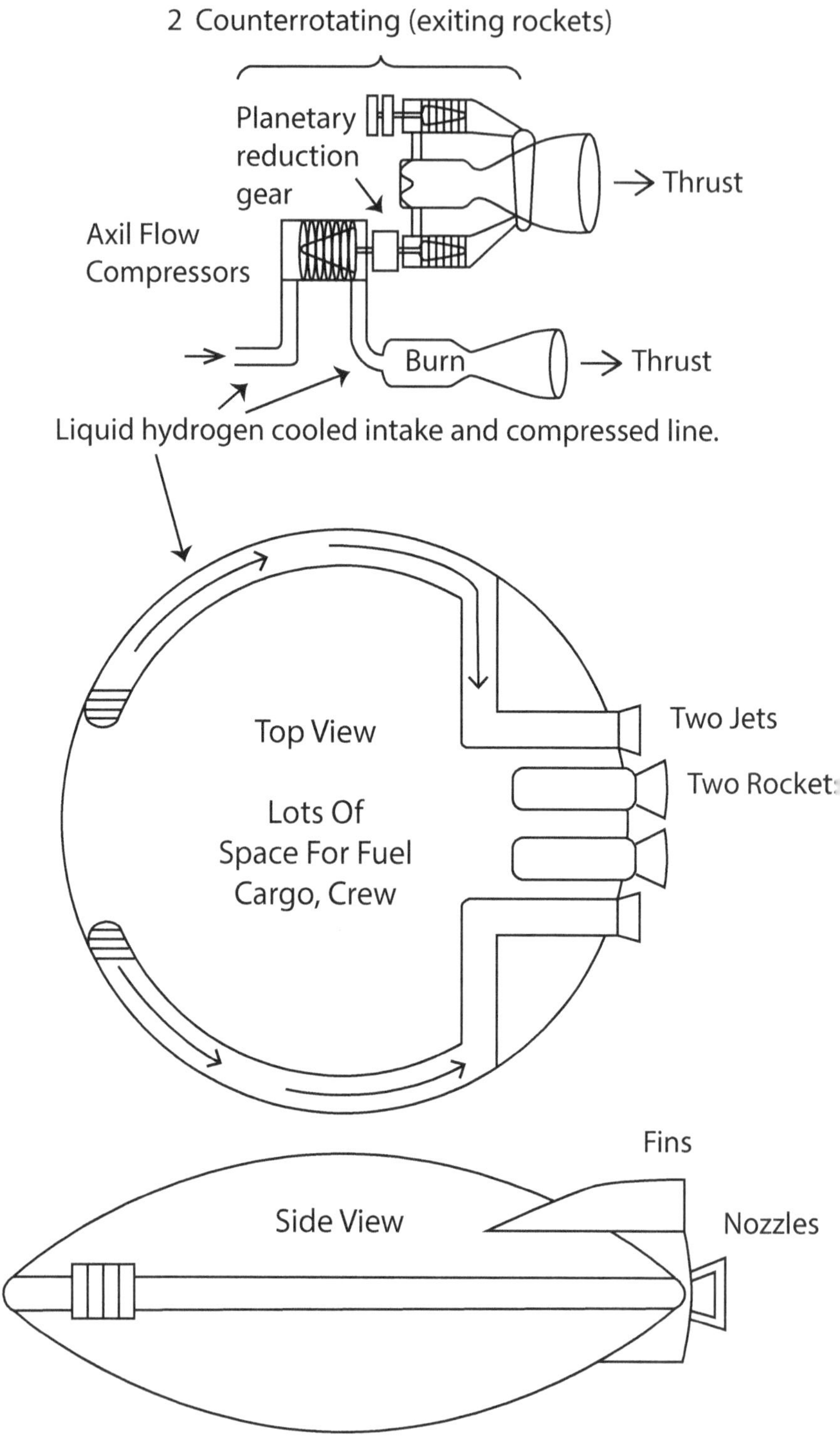

Gear Reduction Flying Saucer

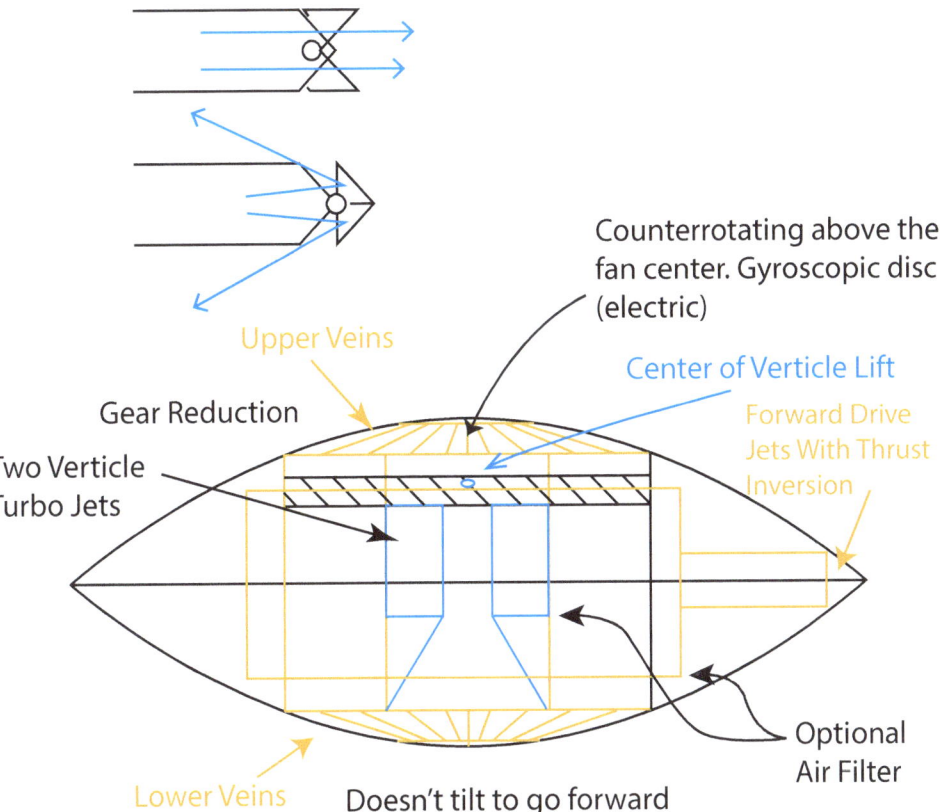

The ship can not flip over accidentally because the center of lift is well above the center of weight. It has a tendency to fly level. Lift exceeds weight because of gear reduction at full throttle. It is not a ground effect vehicle. It is verticle lift with forward drive.

The 1977 UFO Motor - Ionic, Magnetohydrodynamic

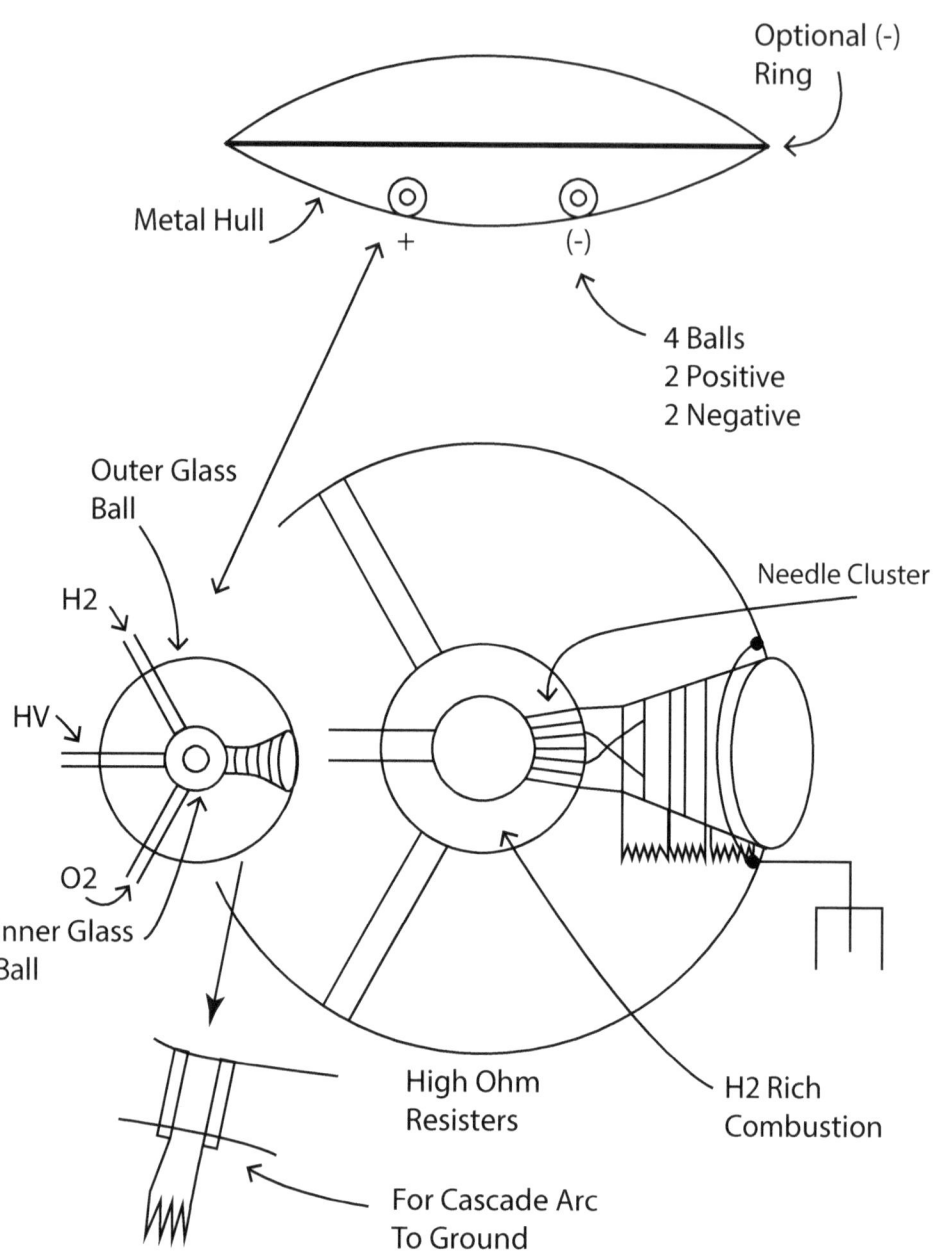

UFO Motor - Transparent Ball

In the year 1973 I saw this description of a UFO in a Gossip Magazine. It was a fable story where the aliens picked up a human. While in the UFO the human asked to see their motor. The aliens showed him this. No explanation was given.

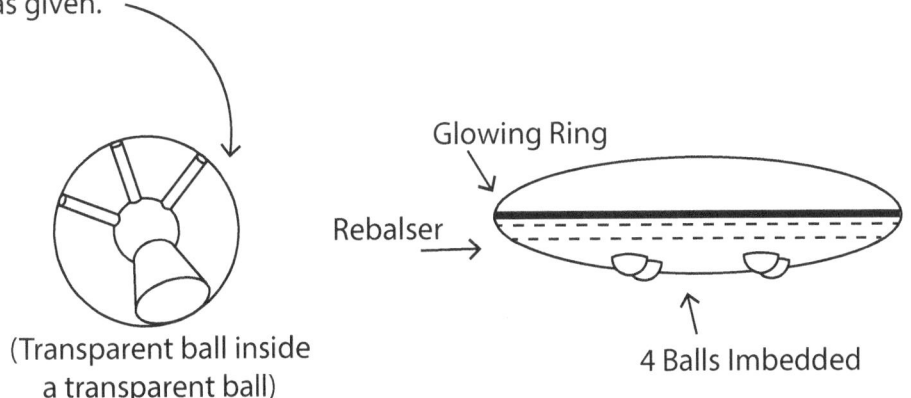

(Transparent ball inside a transparent ball)

4 Balls Imbedded

From the outside the flying saucer in that 40 year old account had a glowing ring but the glow was a corona toward the hull bottom. The glowing ring could have been the emitter (hollow hot ring porous). The rebalser is flush with the hull for vertical lift. The balls are pitch, yaw devices.

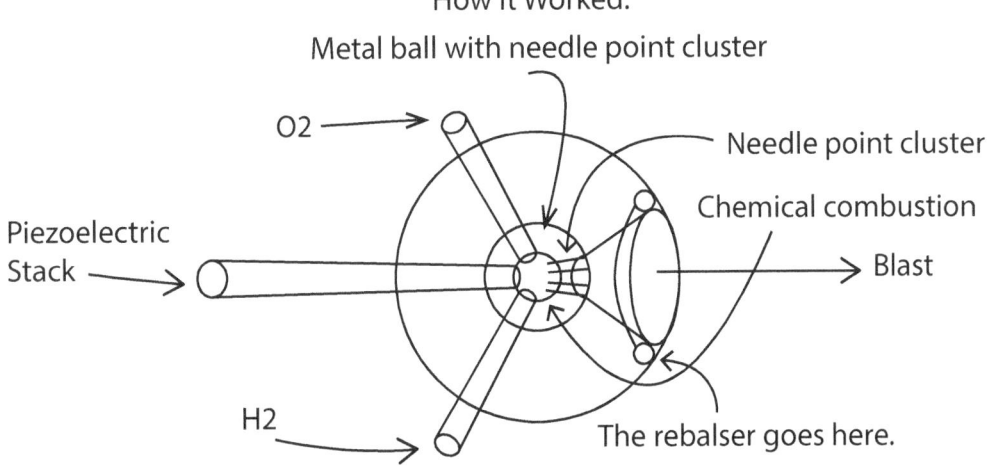

In August of 1973 I went to Washington, D.C. to show NASA this tranparent ball concept except the rebalser was missing! NASA was not interested. I thought then the device was DC. I didn't discover the rebalser until about 10 years ago (2007). I now believe it was actually AC.

Can Transparent Ball Inside a Transparent Ball Generate Thrust? Yes

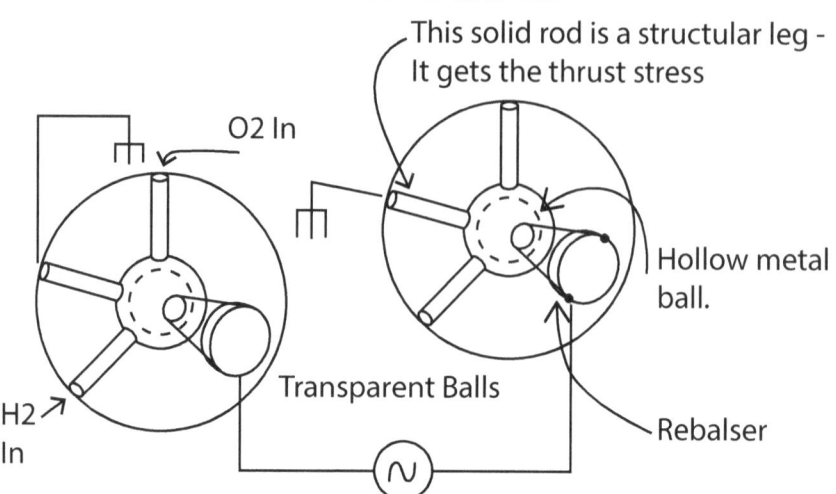

In the hollow metal ball is where some combustion occurs. Ionization at the nude surface, perforated exit lets gass accelerate by electric field attraction using AC wave + then - ions at radio frequencies toward the rebalser. (glass coated collector) The hollow metal ball is glass (foam glass) that is super insulated having a transparent glass glaze for strength.

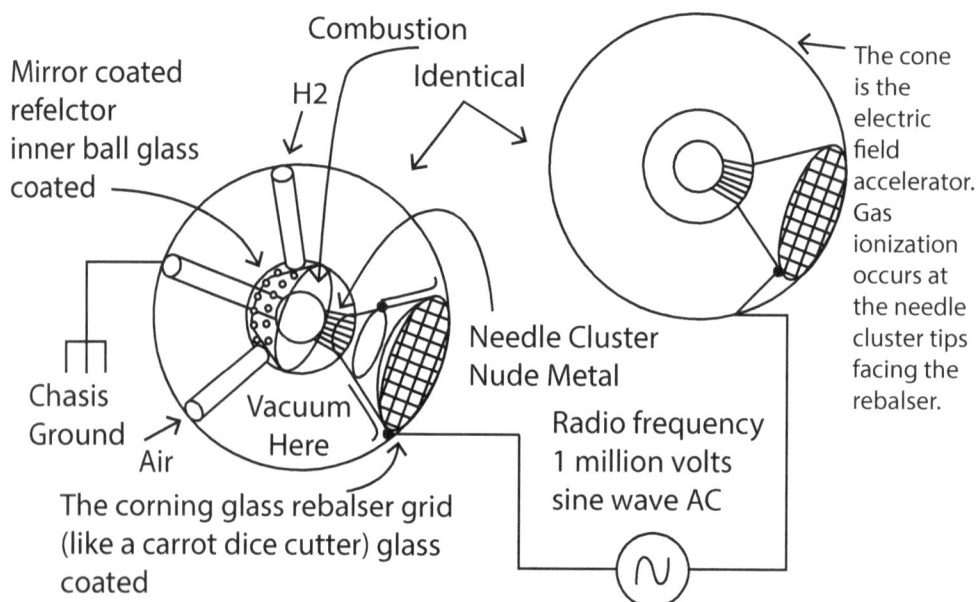

The combustion is not high pressure high temperature but to pre-heat the gas diffusing through the needle cluster. It's the identical design that I created in 1985 then as a DC thruster. A DC thruster requires a nude (exposed metal) collector. The Rebalser is the collector 100% corning glass a such an AC ionic thruster.

What happens if the WC Weirding Field Is Ugraded To Stack? This.

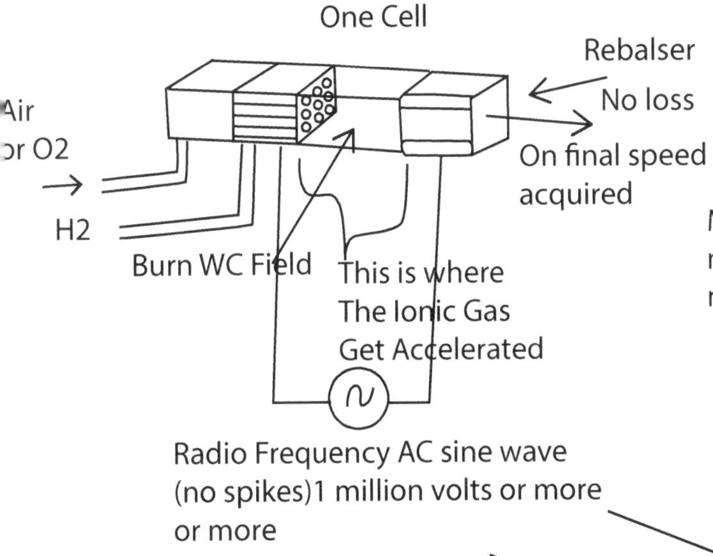

One Cell

Air or O2 →
H2 →

Burn WC Field

Rebalser
← No loss
→ On final speed acquired

This is where The Ionic Gas Get Accelerated

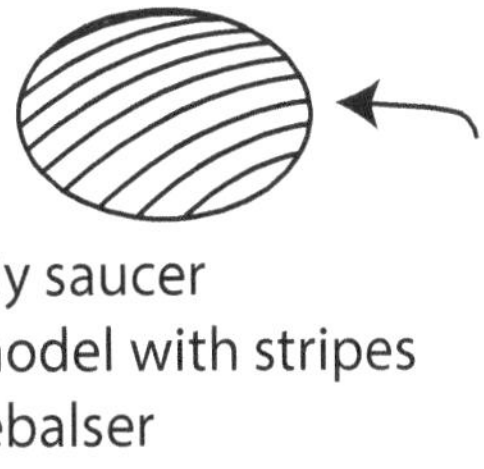

My saucer model with stripes rebalser

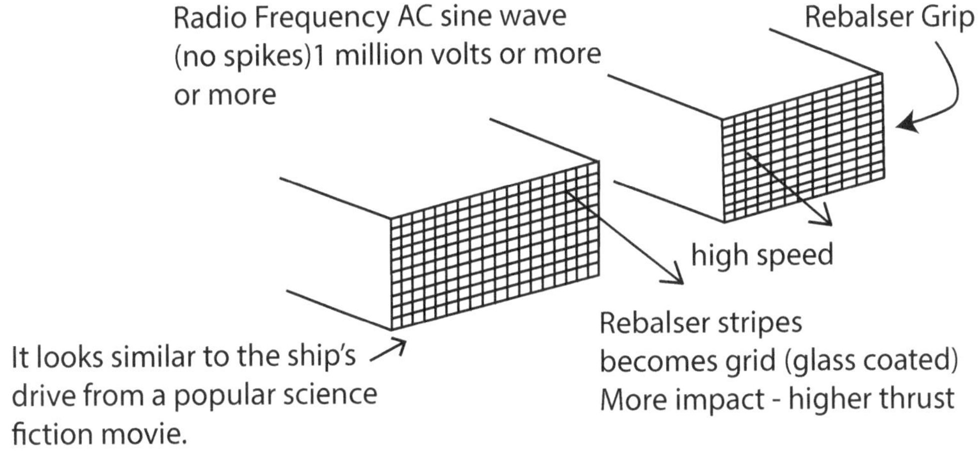

Radio Frequency AC sine wave (no spikes)1 million volts or more or more

Rebalser Grip

high speed

Rebalser stripes becomes grid (glass coated)
More impact - higher thrust

It looks similar to the ship's drive from a popular science fiction movie.

Vacuum ISP <u>Isp astronomical</u> with high Static Thrust In Atmosphere

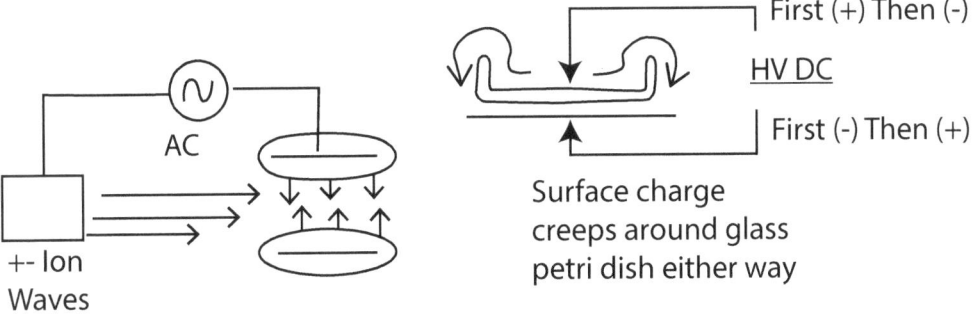

+- Ion Waves
AC

First (+) Then (-)
<u>HV DC</u>
First (-) Then (+)

Surface charge creeps around glass petri dish either way

Rebalser 100% glass insulated results - <u>no hot spots!</u>
The surface charge (creep to green) neutralizes the passing gas without slowing it down - electronic leave glass surface at radio frequencies.

Transparent Ball Motor Upgrade
(Combining Hall Motor With Plasma)

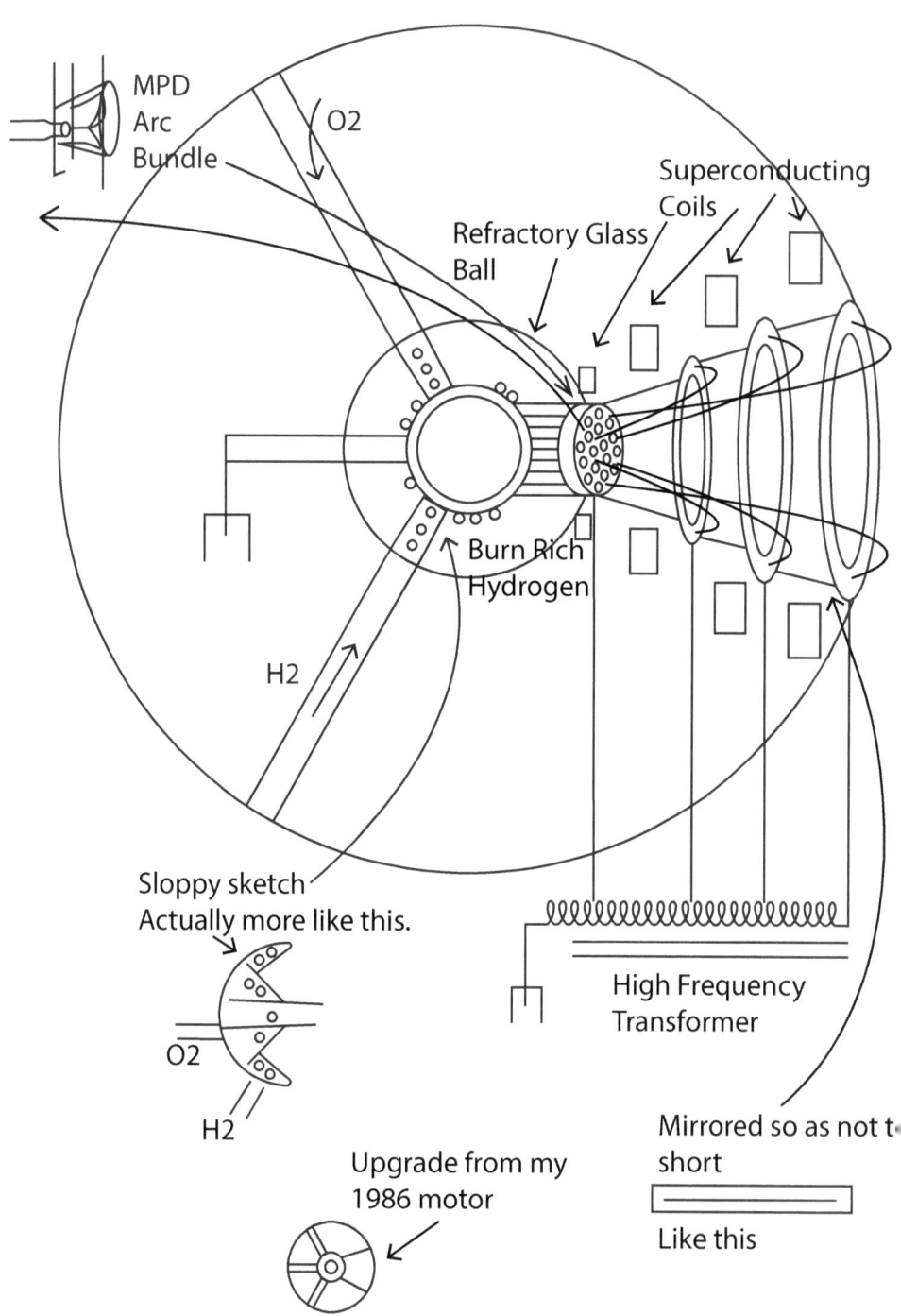

UFO Motor - Glass Ball Swivel

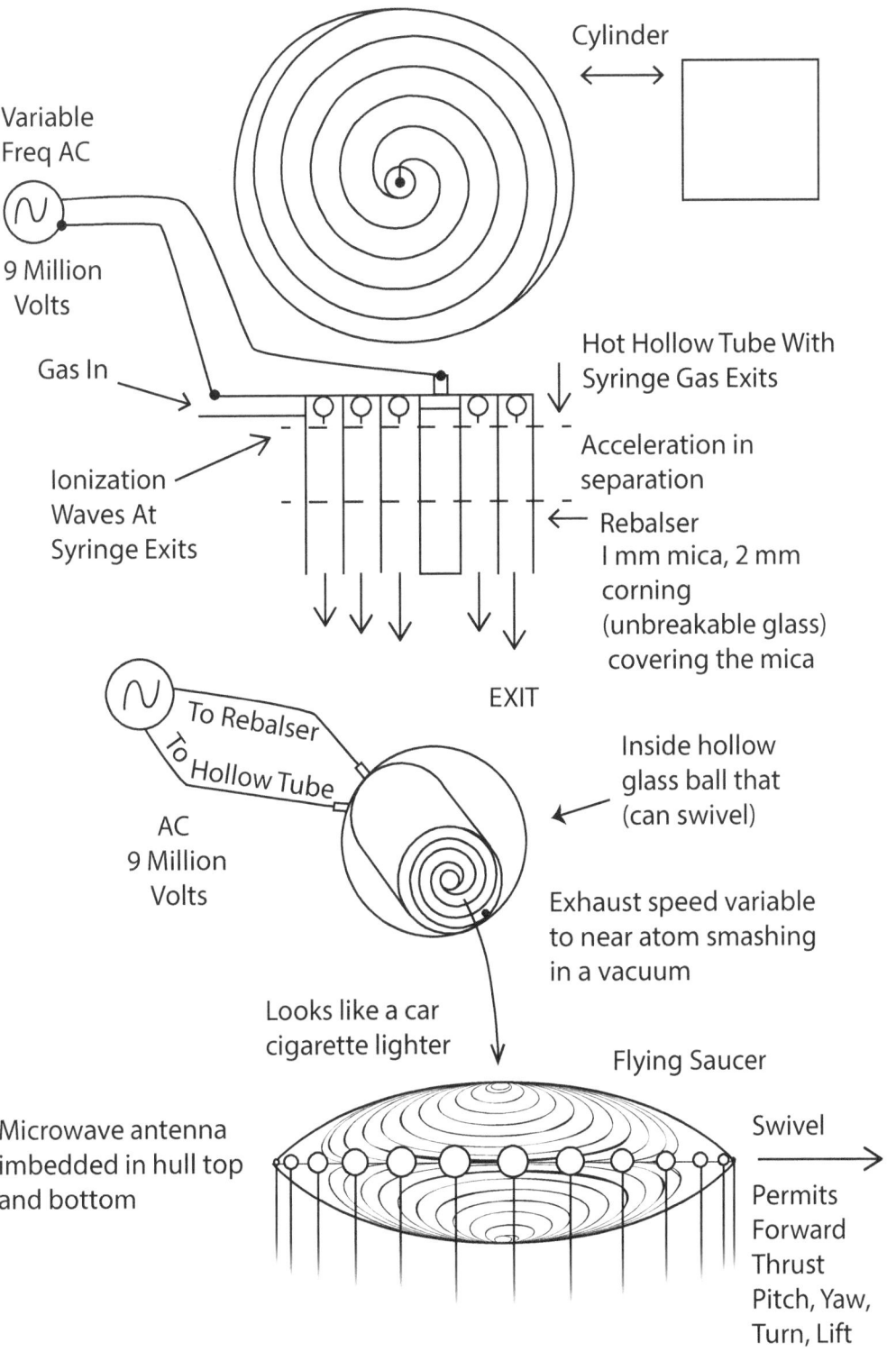

The Spiral AC Ionic Wave Motor

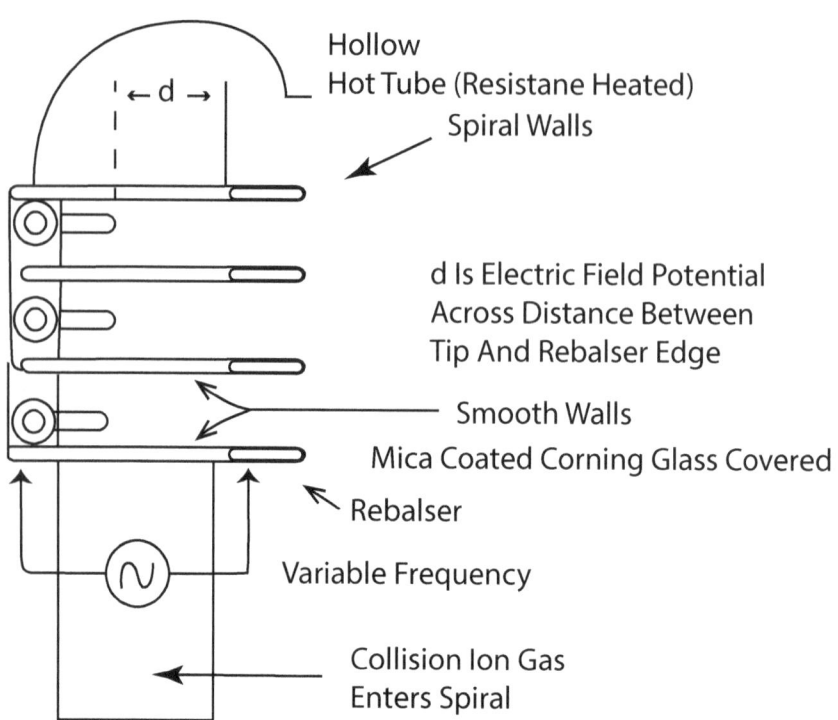
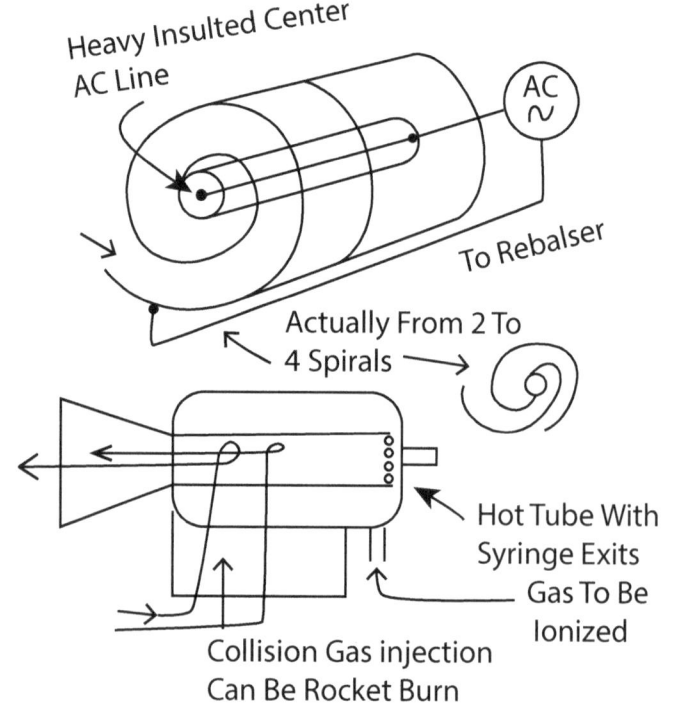

The Spiral AC Ionic Wave Motor

Spiral eliminates crossover electrical contact because at high exhaust speeds would disintegrate. Instead the high ac tension rebalser is by way of common center tap. Any electrical field intensity past 1,000 volts/millimeter ionizes gas at syringe tips. that charged gas gets strongly attracted toward the debalser walls colliding with neutral gasses in the volume between tip and rebalser.

} Like Mighty Mouse With Super Strength pushing A Red Ball Truck

$$F = \left(\frac{Q_1 Q_2}{R^2}\right) K$$

K = Electric Field Conversion Constant

The rebalser does the inverse, it neutralizes the ionized gas by surface charge AC. Gas exit speed is $z \sqrt{\frac{2 Ep}{M}}$ In Meters/Sec

M = Aberage Mass (Ionized Gas And Neutral) Affected

Ep = 1/2 Mv² Kinetic Energy } Related to electric power P = IE
V = Meters/sec (Not Volts)   supplied.   I = Current
                                                           E = Volts

The Spiral Concept - Upgrade

This subjects combustion gas to avalanche collision.

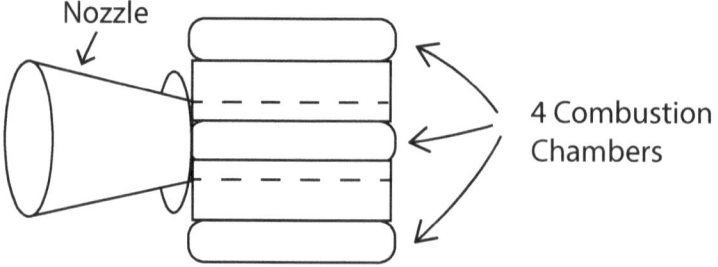

Notice The Clear Unobstructed Exit

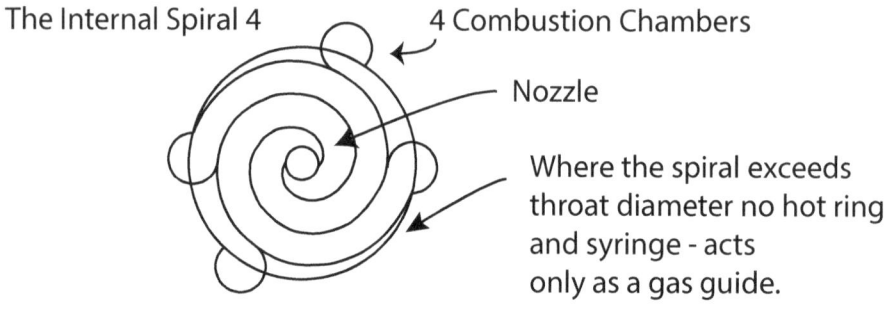

Where the spiral exceeds throat diameter no hot ring and syringe - acts only as a gas guide.

The Chemical Burn Enters The Spirals And Exits Out The Nozzle

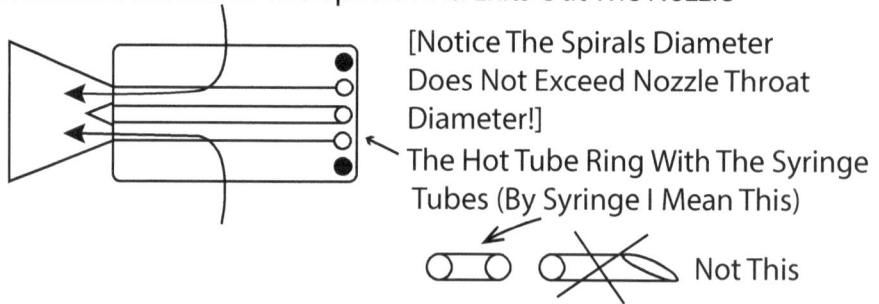

[Notice The Spirals Diameter Does Not Exceed Nozzle Throat Diameter!]

The Hot Tube Ring With The Syringe Tubes (By Syringe I Mean This)

The resistance heated emitter is surrounded by super insulator (like the space shuttle tile substance) non conductor

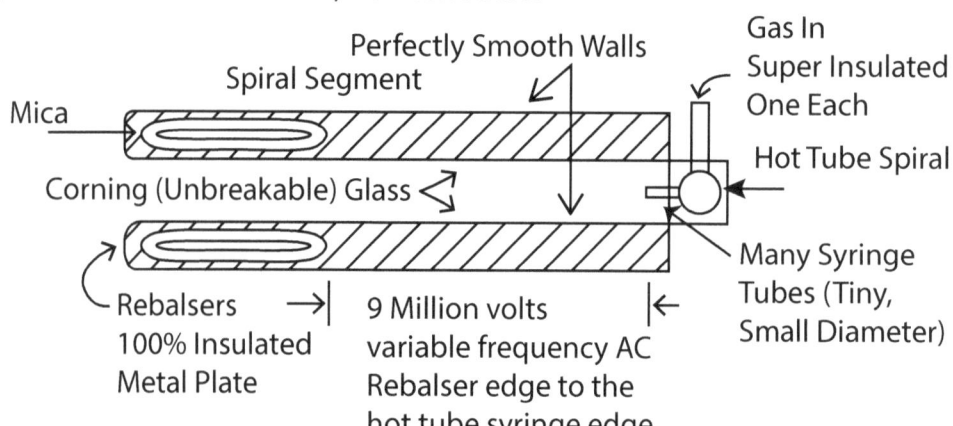

With Ultraviolet Laser Modification Added: The AC Ionic Wave Rocket

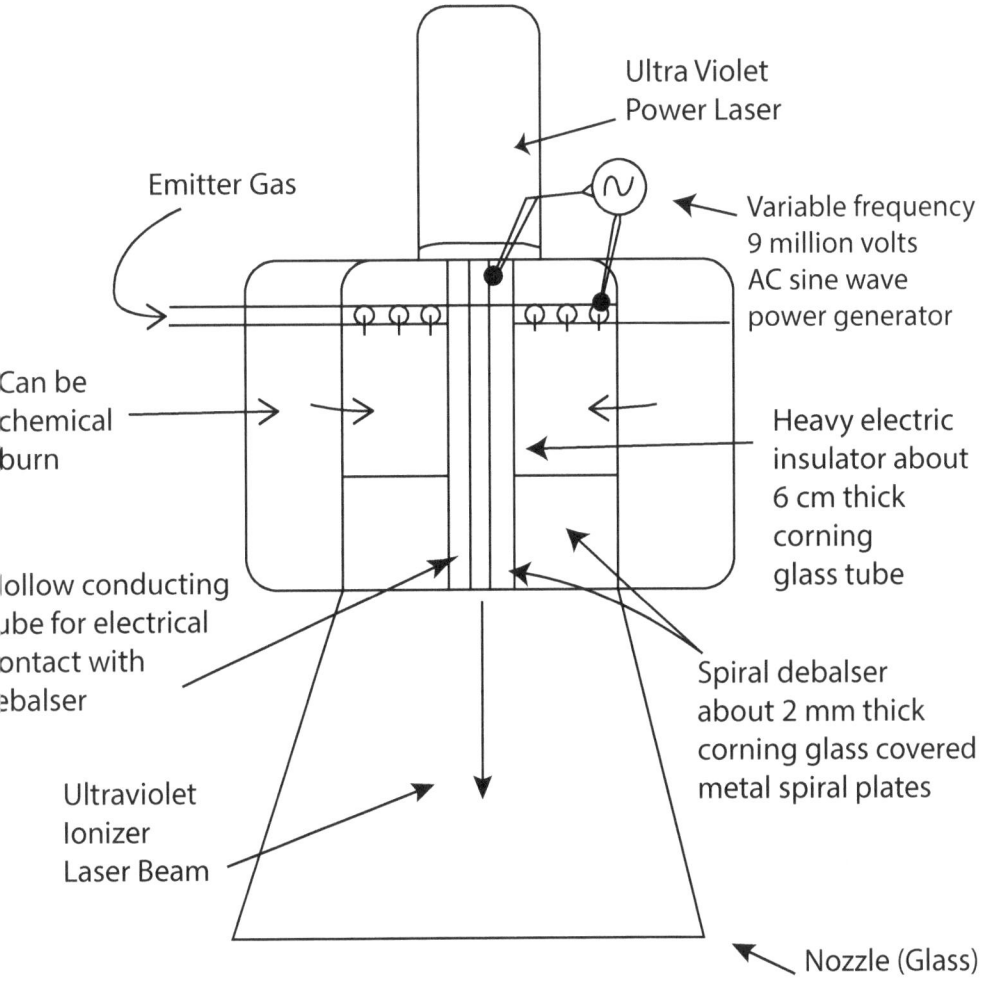

The ultra violet laser acts like a nude metal rod with high resistance. This causes intensified + - ionic wave flow from the emitter past the rebalser to the exhaust in the nozzle. The electric polarity is the same as the rebalser spiral polarity but opposite the emitter syringe polarity. It's like the equivalent of a much larger rebalser. Notice because the rebalser is the same polarity that there is no release flow, hence it acts like a triode tube grid that intensifies syringe emitter ionization.

Permanent Dielectric UFO Power Generator

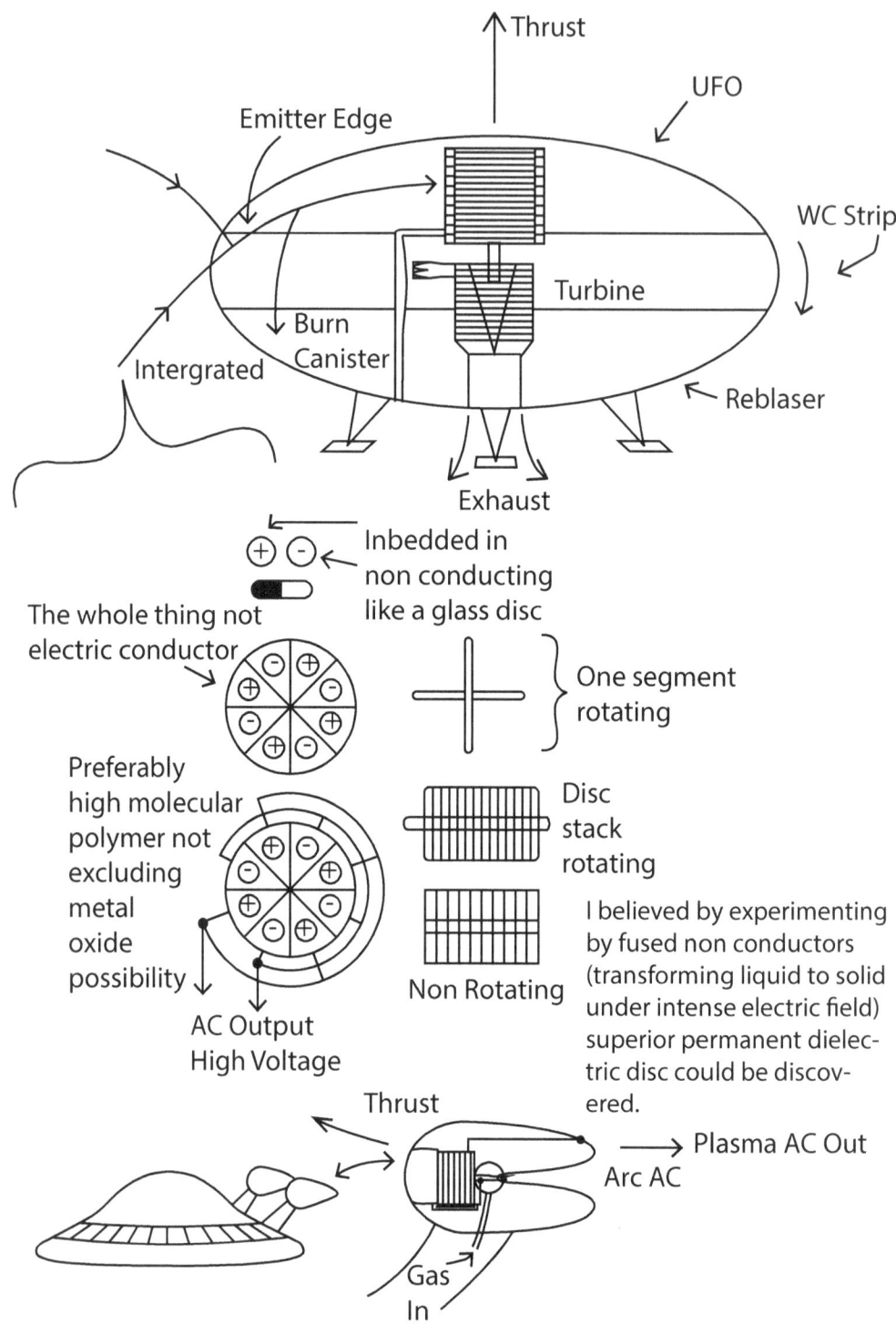

The Possible Tesla UFO Motor of 1920

The Possible Tesla UFO Motor of 1920.
Assuming that he had conducted this experiment.

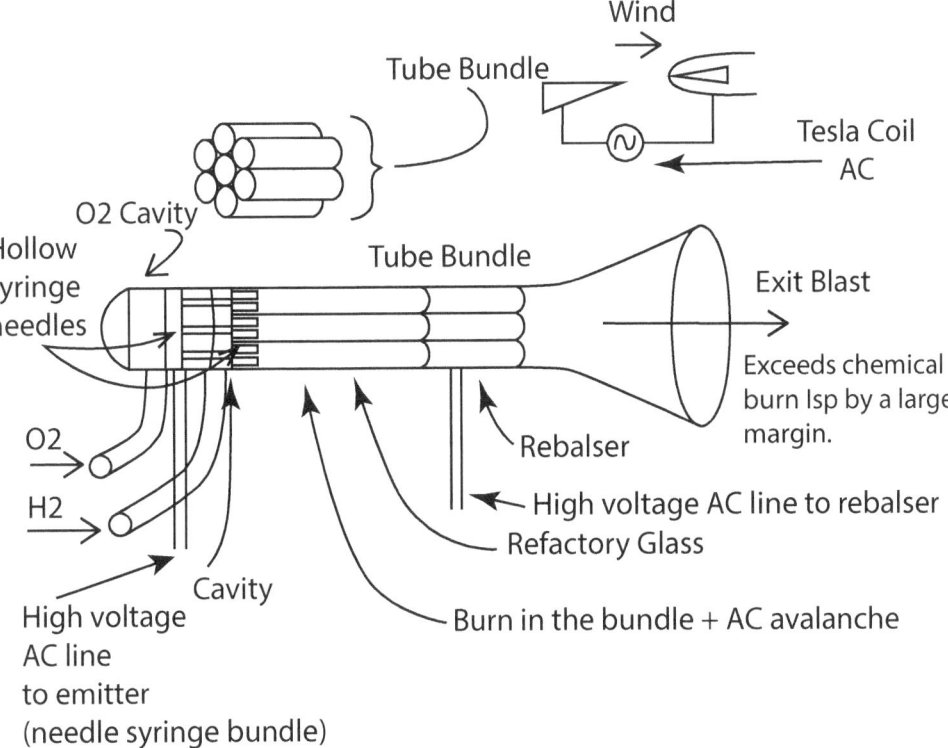

In one motor there is both chemical burn and <u>AC</u> ionic wave acceleration avalanche high thrust, plus high exhaust speed that exceeds a chemical burn (only) rocket - both in the atmophere and a vacuum (space).

To Mars And Back Without Stages

The rocket pumps drive the Nikola Tesla rocket electric generators. After velocity or orbital the nuclear electric drives a smaller number of Nikola Tesla rockets.

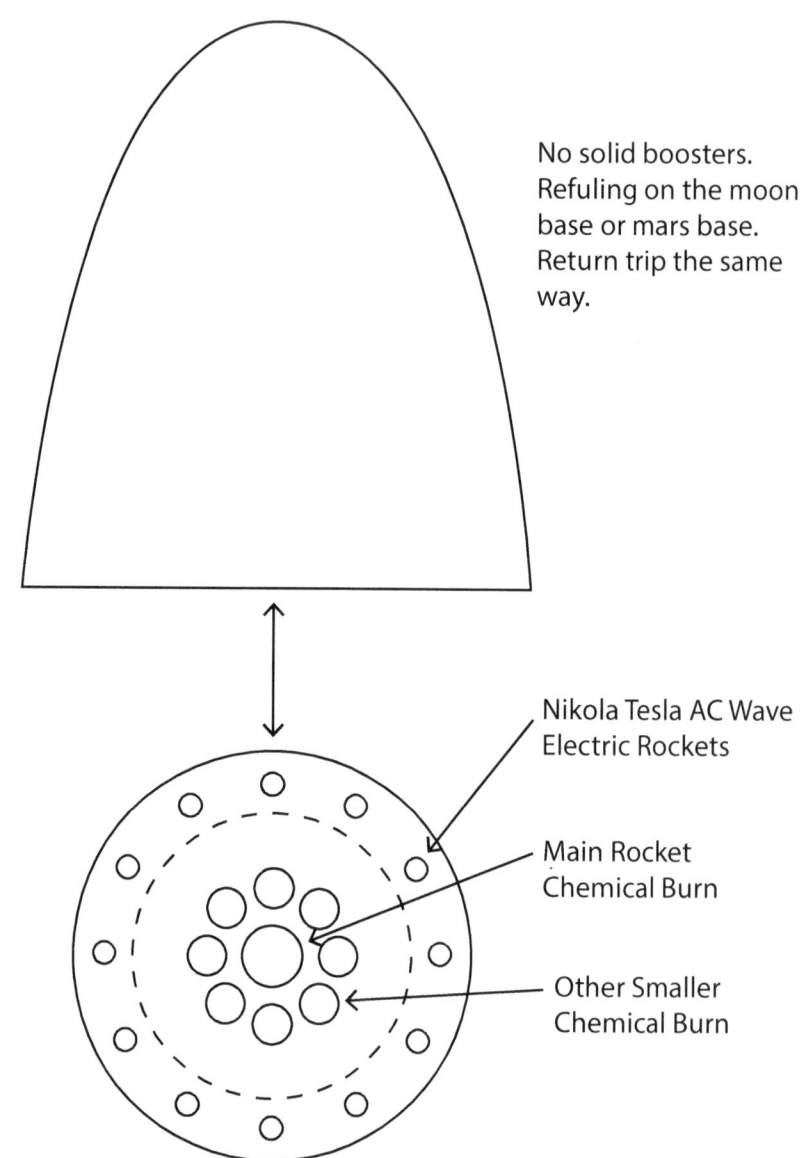

No solid boosters. Refuling on the moon base or mars base. Return trip the same way.

Nikola Tesla AC Wave Electric Rockets

Main Rocket Chemical Burn

Other Smaller Chemical Burn

Possible Nikola Tesla Motor (Upgrade)

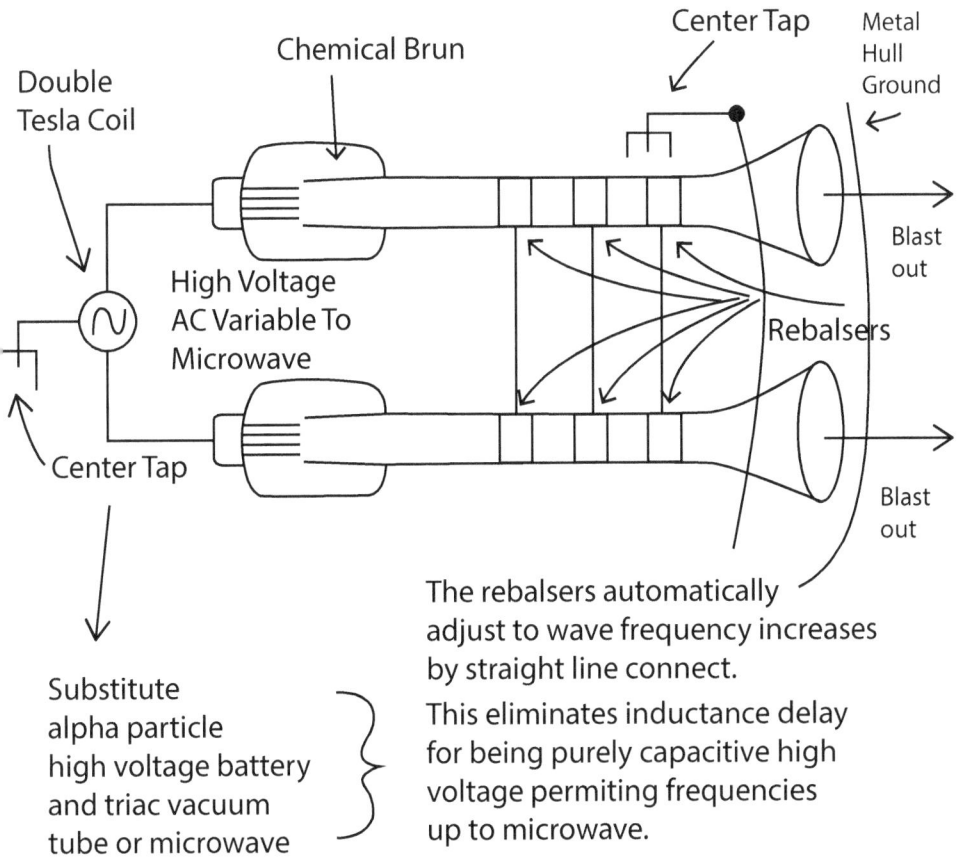

Nikola Tesla's Blimp Thruster - 1,000,000 Volts

Cone needle point to a tube
Needle points part ways in each glass tube.

Gas in
Hollow Metal Ball
Hull ground

|← 10 cm →|
Glass
Nude metal tube here

Hollow Glass Bundle
AC High Frequency
Rebalser here

Lightening arc

Glass
High voltage AC or DC
Hull Ground

(About 10 cm length)

Nikola Tesla's "blimp" thruster could have been as simple as this.
At 1 million volts tube length 4 meter (from ball to nude tube).

Did Nikola Tesla Invent the Rebalser?

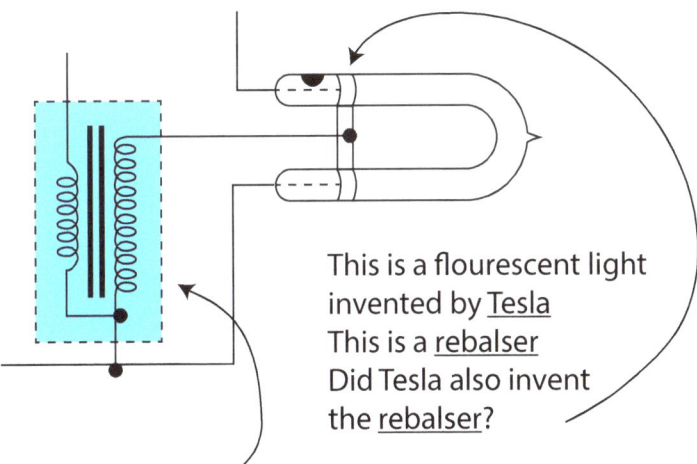

This is a flourescent light invented by Tesla
This is a rebalser
Did Tesla also invent the rebalser?

Varying R, pot changes the frequency of a strobe flash.
This is a Tesla coil step up transformer powered iron core.

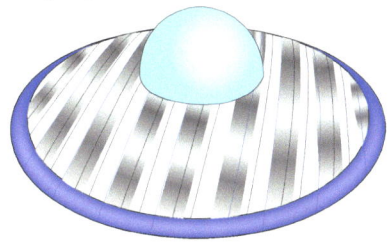

    I already said what that black strip is. Remember the American Flag? Rebalser is Italian for bounce like a ball bounces from a wall. In this case the balls are waves of ionized positive followed by negative covalent bond gases in rapid succession with the electrons leaving the rebalser surface to neutralize the plus molecules without slowing down hence rebalser. When the molecules are negative the electrons leave the gas molecules for having passed so close to the rebalser surface. Like a shaving razer cuts your whiskers only while passing close to the face. It has to be AC otherwise the rebalser half capacitor will fully charge and that's stopping the motion of a shaving razor on your skin while shaving. That's why DC won't work the way this motor is designed. The horizontal WC is for vertical lift. I don't need it. The horizontal WC suffices to demonstrate the motor works. With no funding that's the best I can do. Aliens with a big lead have a variety of advanced propulsion motors. Be content I have one that works. You see it's hard for a fiction writer to write 100 percent fiction. Without realizing it some UFO writers intending to fool accidentally told a few true lies without meaning to spill the beans.

If 1 ampere charge is separated by meter distance

1 meter

the force of attraction can lift a battleship out of the water. In a circuit 1 ampere takes about 1 hour to travel 1 meter because flow is in parallel - However at the needle points the electron flow is close to single file - but the magnetic field per meter length is the same.

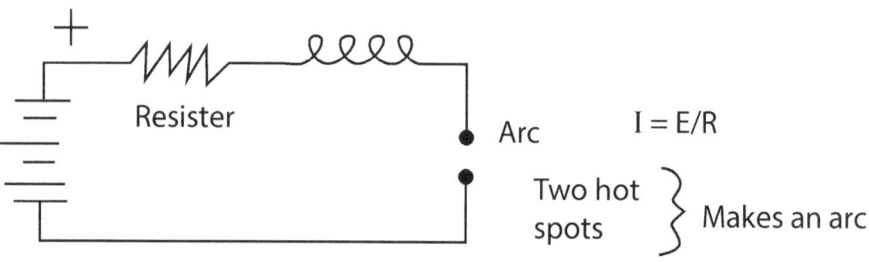

So then what happens with only one hot spot? For the magnetic field to read the same intensity the ions single file have an immense speed - like cosmic rays creating an avalance collision - Hence the flow is in the picofarads of charged ions. The amount of energy it takes to separate one ampere charge one meter from the opposite charge is the energy needed to lift a battleship out of the water by a strong cable. I am just using an infinitesimal fraction of that energy which is why my motor has a weak thrust. To increase the power requires raising the voltage and increasing the picofarads current flow as corona. That capacitor is in picofarads. How much charge separation it stores makes my corona motors generate the trust you saw.

The WC Strip Discovery

A transparent plastic tube held by my bare hand close to Van de Graaff.

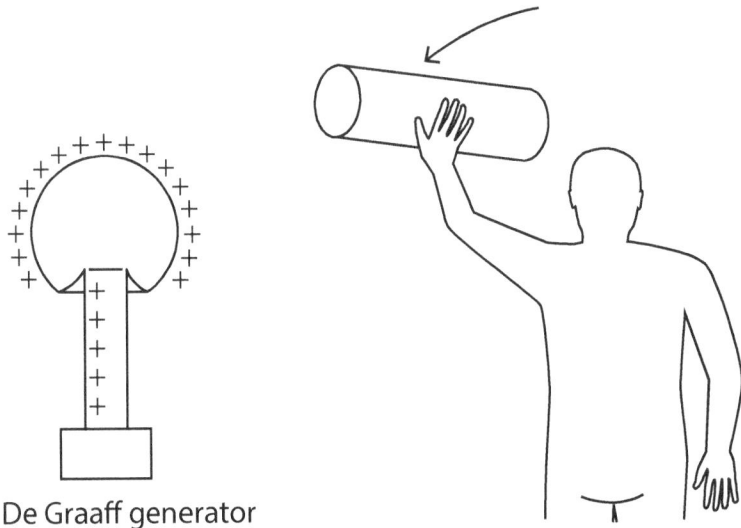

Van De Graaff generator

In the year 1975 I conducted an experiment where I moved away from a Van De Graaff holding a transparent plastic tube. I then stuck my hand inside the tube and got literally knocked off my feet.

What happened?

In all typical lightning bolts -

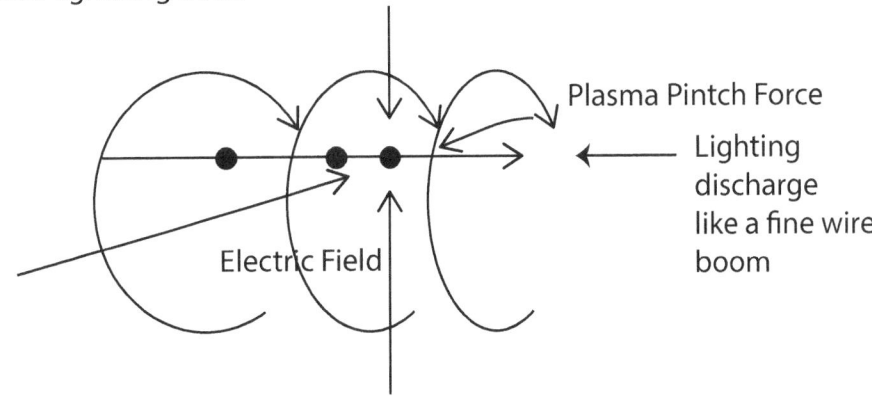

That's not the kind of boom that knocked me down. What knocked me down was the equivalent of getting too close to the WC strip of a corona like in my AC wave flying saucer model experiments.

[The AC Ionic Wave Thruster Is Corona - Not Lightening Bolt]

In a solid conductor the flow speed is about one meter/ hour. In the corona; the flow speed is very fast - can approach near light speed at elevated voltages and near vacuum in the tube.

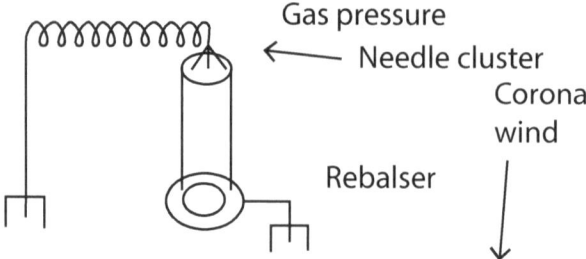

In every case; the magnetic field measures the same. But why? Because in a solid wire the flow is many charges in parallel. In the tube needle point exit - In series (less parallel) but in one second the number * speed (solid wire) = number * speed in a near vacuum $I_A V_A = I_B VB$. If the gas pressure is raised to one atmosphere - the electric field intensity gets raised to one thousand volts/mm. Why? Because the electric field attraction results in a great many collisions with gas pressure rise. What the rebalser does is prevent a hot spot from occuring. Hence:

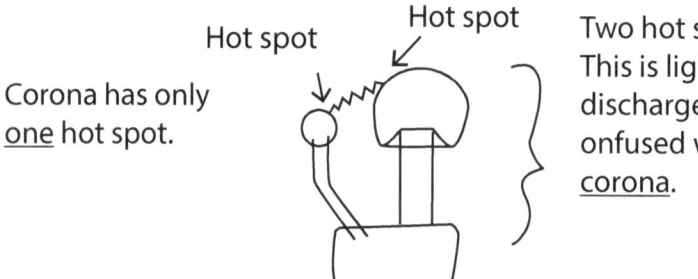

Corona has only one hot spot.

Two hot spots (no wind) This is lightening discharge not to be c onfused with corona.

The phenomena is purely corona, whereas in the van de Graaff it is lightening discharge. So then the AC ionic motor is a corona motor. Very closely related to the phenomena of lightening bolts in a cloud firing vertically into the ionosphere. Photographed by the international space station astronauts as a corona wind.

How AC Differs From DC

To the eye AC and DC looks the same.

How AC differs from DC.

AC results in a reblaser. In DC the rebalser is exposed nude metal that shorts - will not work except in a very low thruster.

Porous AC <u>emitter</u> is not a rebalser.

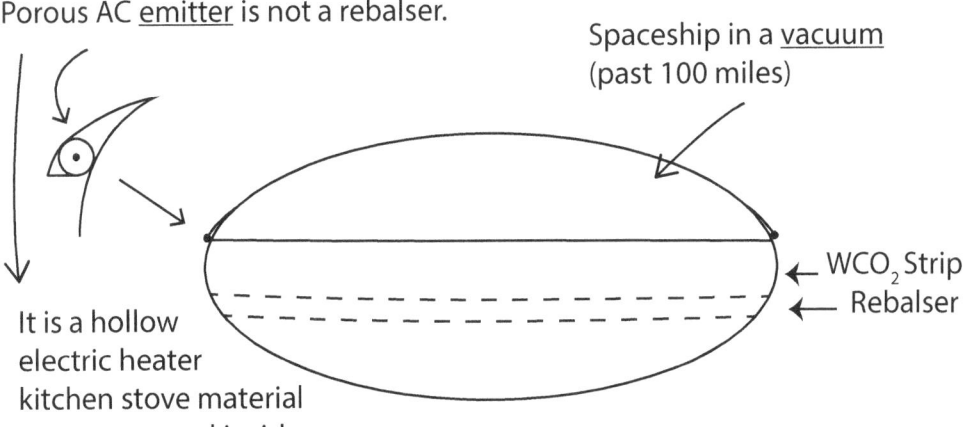

It is a hollow electric heater kitchen stove material so gas pumped inside can leak out - It doesn't rust. When on it glows red hot. The gas on the side facing the rebalser gets transformed into negative ion molecules when negative high, followed by positive ion molecules, when positive high in AC high frequency. Unlink DC that can arc in a short circuit, because the rebalser is a coated insulator - no exposed metal.

So What?

The formula $Pe = MV^2$ doesn't give a darn if the mass = M drops to almost nothing.

V = speed of the charged ion molecule across the WCO^2 strip.

So if the gas, input is low, the thrust is the weight of a sheet of paper. If the gas supplied is high, the spaceship experiences a high acceleration - unlike a DC device that would arc a short and do nothing.

The Flying Saucer Hover Concept

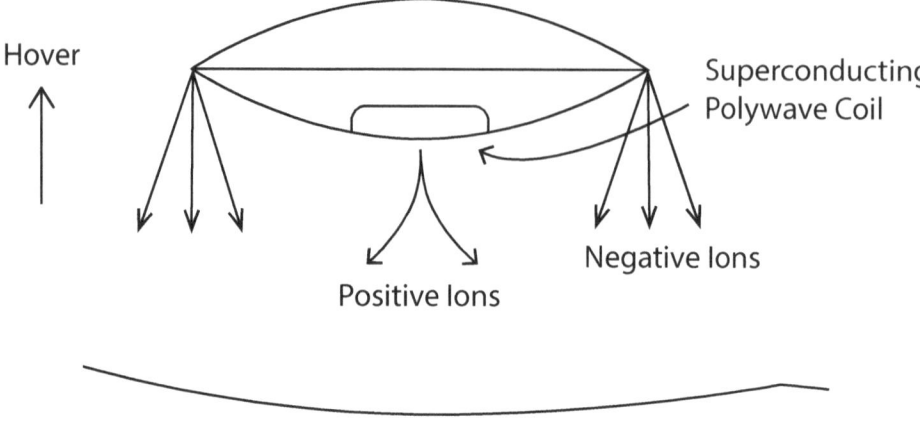

Hover

Superconducting Polywave Coil

Negative Ions

Positive Ions

Common Ground

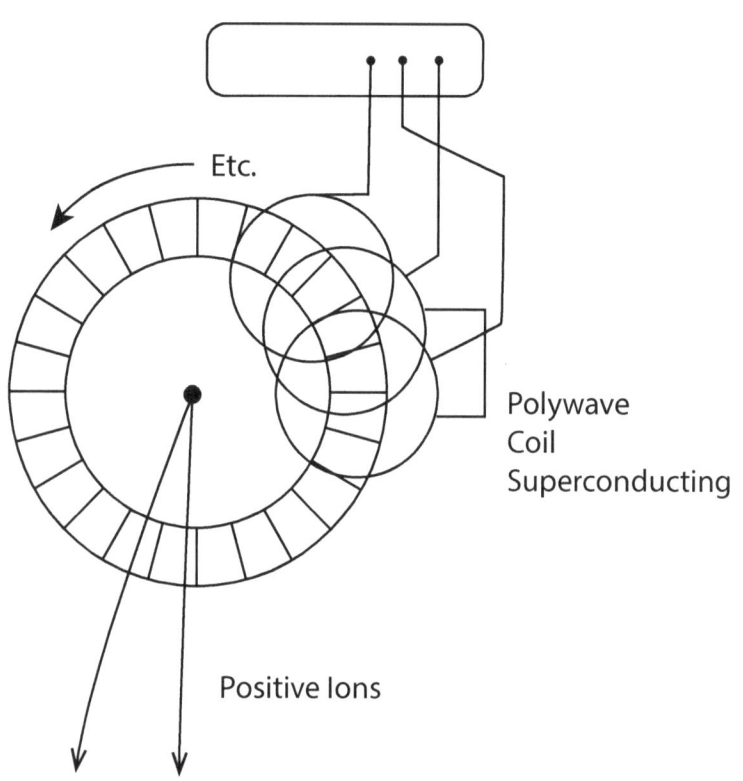

Etc.

Polywave Coil Superconducting

Positive Ions

The Magnetic Field Repels, Interacts
To Push The Saucer Away From The Ground
If The Traveling Wave (Polywave Spins Fast)
(Radio Frequencies)

What Is Antigravity?

You could say that anti-gravity is when mass exhausrt speed at the rocket nozzle exit exceeds the value:

$$\sqrt{\frac{2GM}{R}}$$

For the earth that value is 11.9 km/sec. Otherwise you have a grasshopper - a hop and back down. NASA achies anti-gravity by using chemical burn rockets by <u>staging</u> to orbit.

Gay-Lussac's law: Top side only the pourus exit pacing the rebalser <u>nude</u> and hot. The neutral gas when exiting the hot pourus slot gets an electron added or robbed from the electron shell (atomic). This makes the electron field wet with ions.

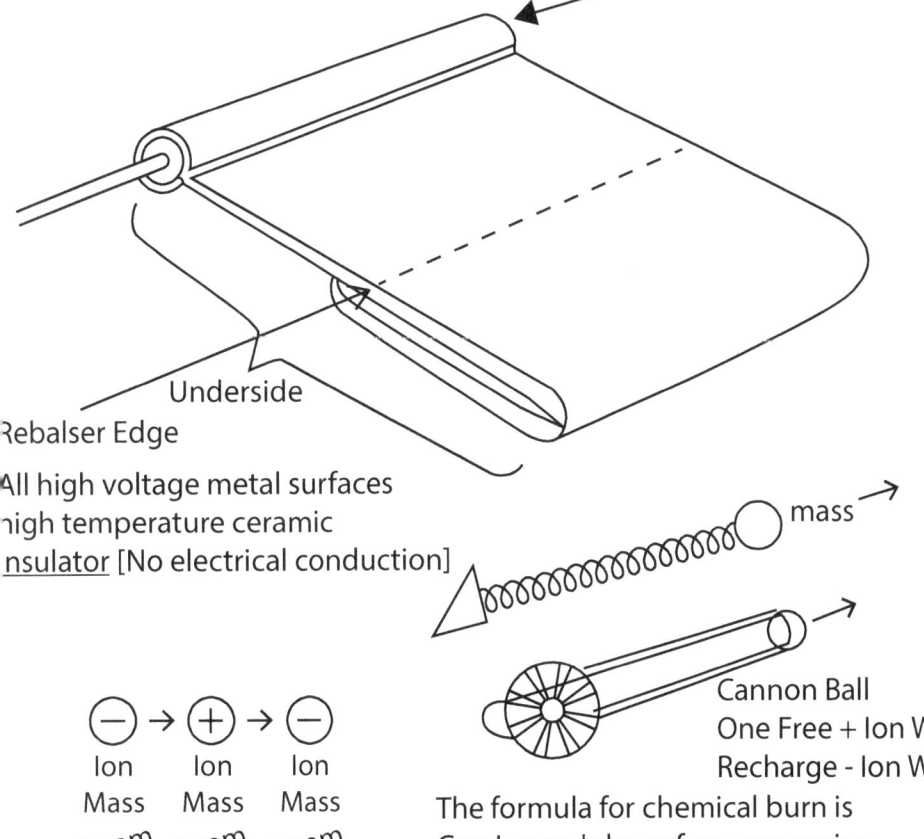

Underside

Rebalser Edge

All high voltage metal surfaces
high temperature ceramic
insulator [No electrical conduction]

mass

⊖ → ⊕ → ⊖
Ion   Ion   Ion
Mass Mass Mass
Boom Boom Boom

Cannon Ball
One Free + Ion Wave
Recharge - Ion Wave

The formula for chemical burn is
Gay-Lussac's law of gas expansion
Einstein CERN Physics { the law for charged mass acceleration in electric field particle physics & accelerator }

Understanding Gravity - Inertia and Kinetic Energy

The story of the graviton is false.
A particle is an electromagnetic feild wad - <u>Everything is light</u>
(Everything material)

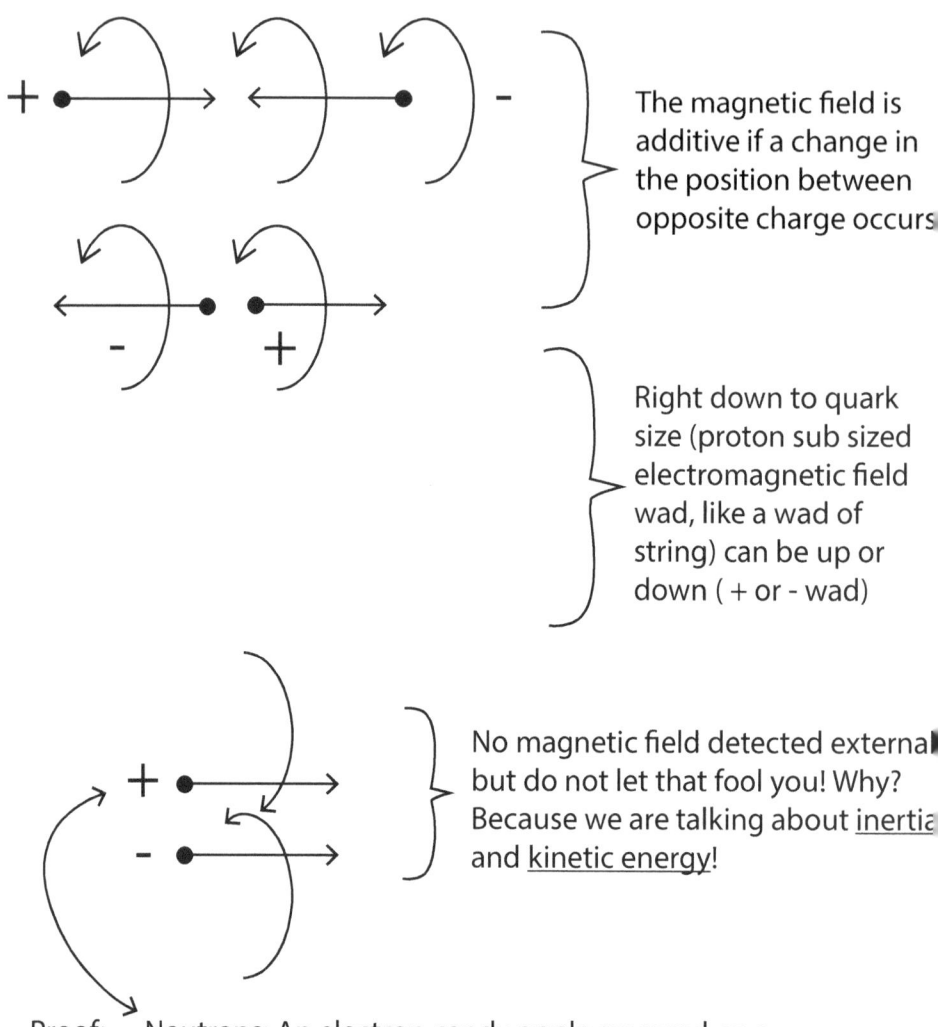

The magnetic field is additive if a change in the position between opposite charge occurs

Right down to quark size (proton sub sized electromagnetic field wad, like a wad of string) can be up or down (+ or - wad)

No magnetic field detected external but do not let that fool you! Why? Because we are talking about <u>inertia</u> and <u>kinetic energy</u>!

Proof: Neutrons: An electron candy apple smeared on a proton! Rule not violated.

- An electon absorbs electromagnetic wave to <u>accelerate</u>. And gives up electromagnetic wave to decelerate! Same for a proton! What about gravity To accelerate or decelerate from gravity the rule is the same! The energy transfers across the gravity field.

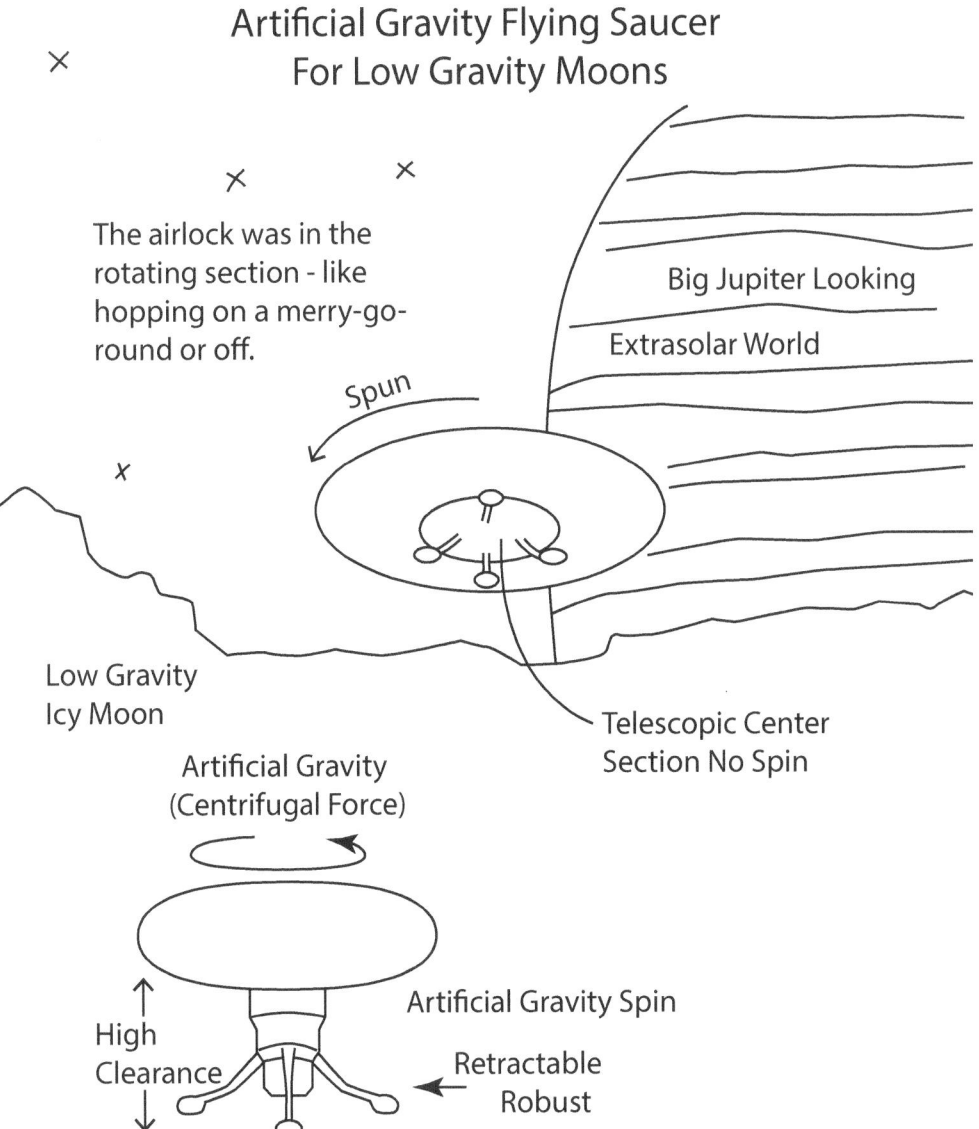

Gravity Is A Monopolar Attractive Energy Field
(A Created Entity).
It Can Be Thought Of As A Standing Wave

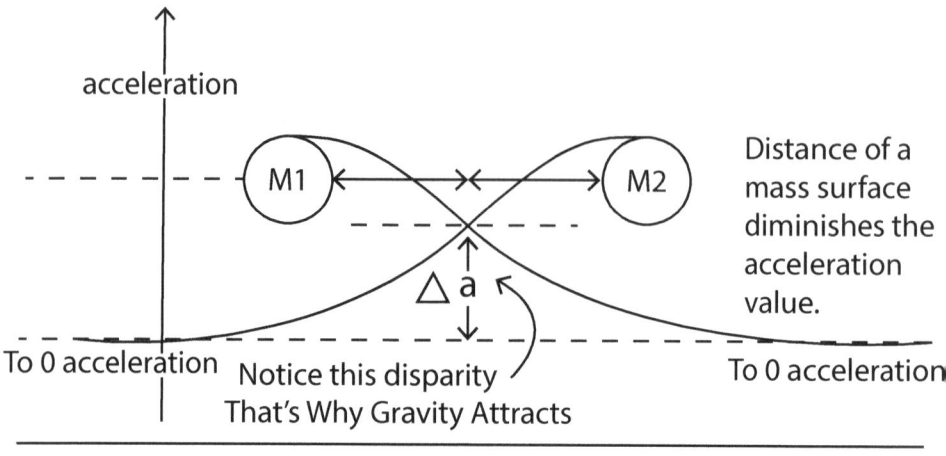

Gravity surrounds any mass even as small as one plank constant. That's why light blends around a big mass but the mass has to be very big or massive to be noticeable. Space curvature is a misnomer for the curvature around the mass where the acceleration value reads the same. It can be illustrated by:

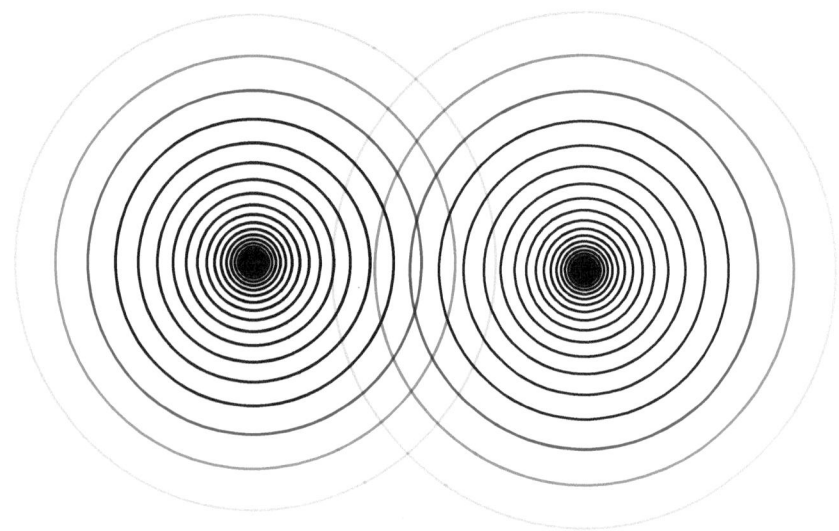

The more intense around masses there is attractive overlap. In the illustration the darker area is more intense, then fading to transparent with distance.

The Bell Motor As A Electric Power Generator.

Bell refers to a Nuclear Fusion experiment that occured in California.

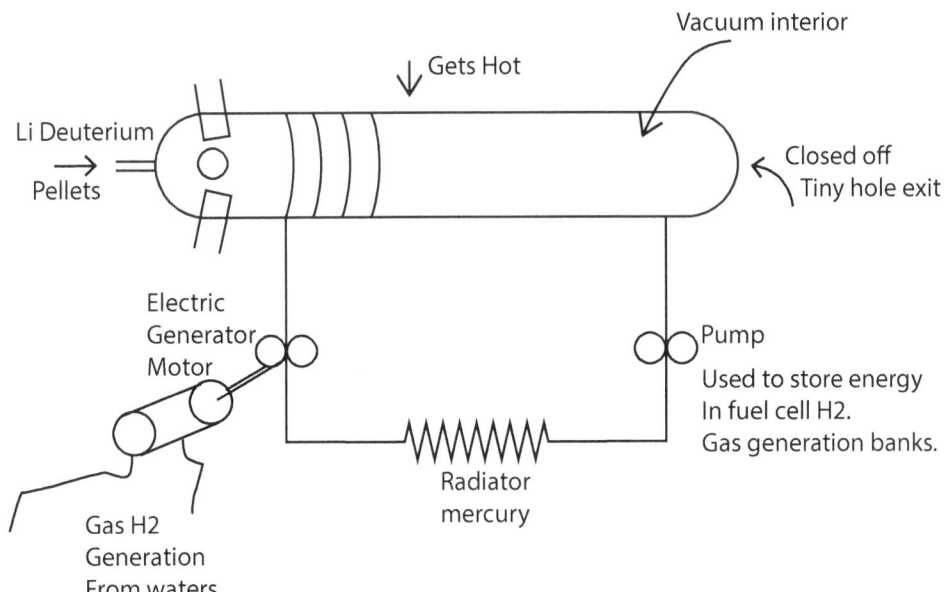

How it works - Very slow. One blast then about an hour later another blast. It supplies power to the spaceship. The Bell thruster another device.

Dallas Texas Public Library
(science news) early 1984

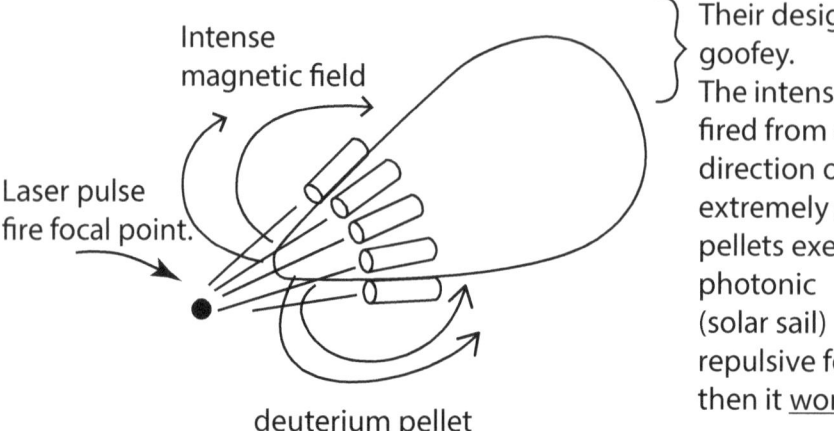

Intense magnetic field

Laser pulse fire focal point.

deuterium pellet

Their design was goofey.
The intense laser fired from one direction on an extremely small pellets exert a photonic (solar sail) repulsive force then it <u>wont work</u>.

So right away I figured this design.

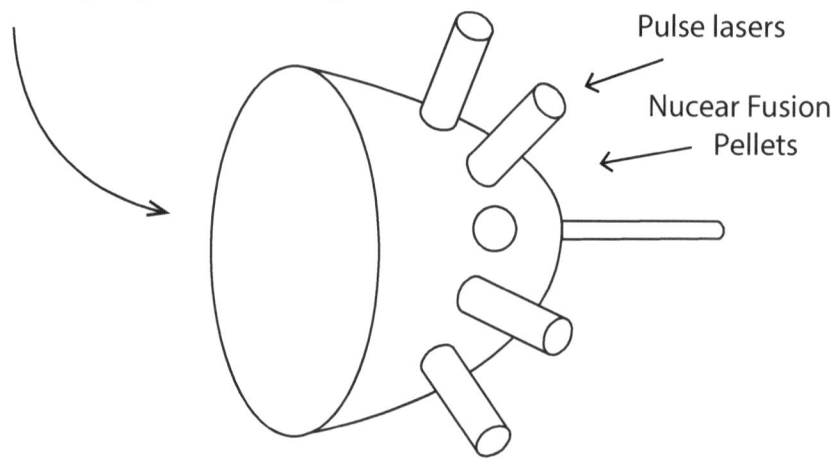

Pulse lasers

Nucear Fusion Pellets

Magnetic field not <u>needed</u>

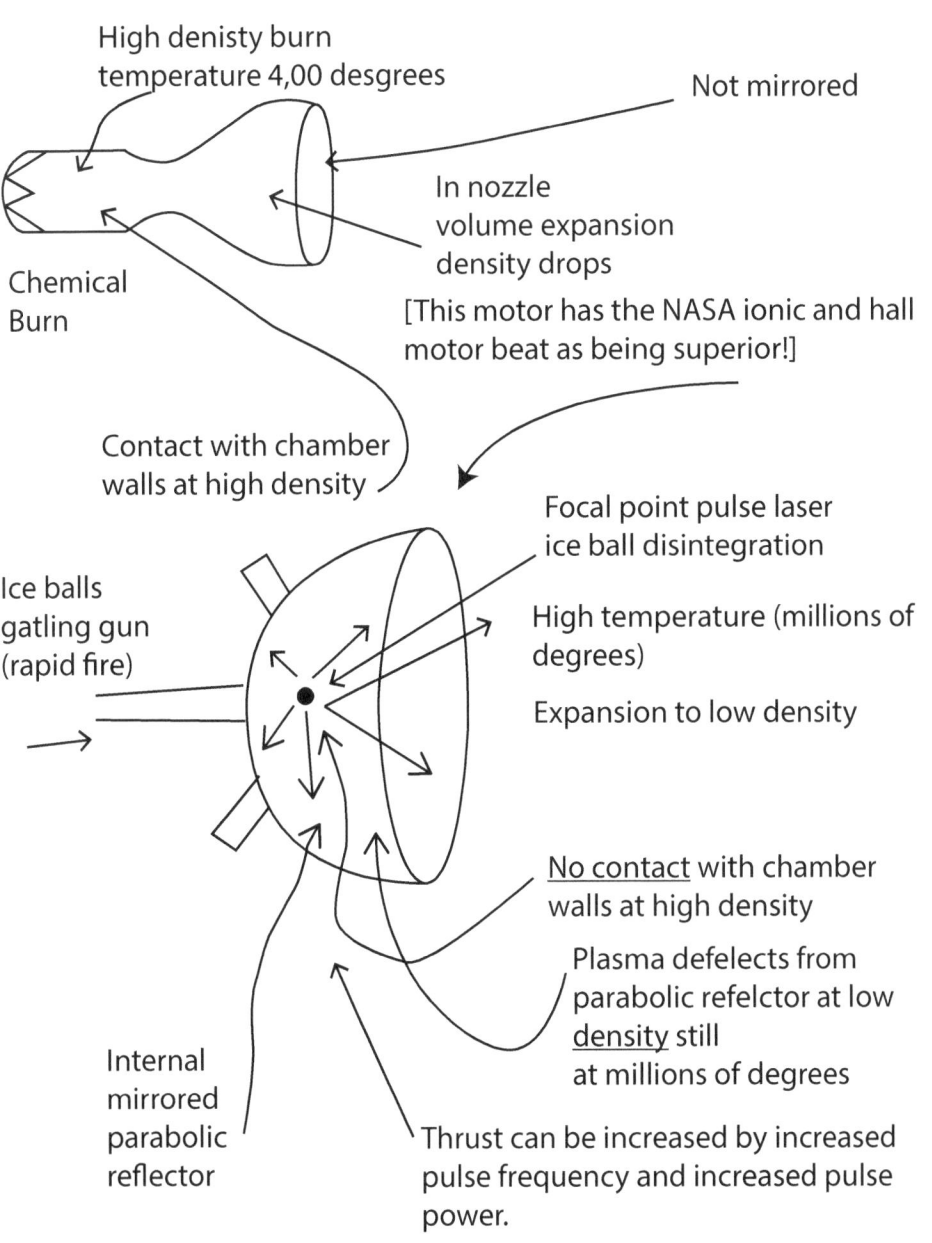

Artificial Gravity In Space

Artificial gravity (spin) diminish vertigo by icreasing radius from center of spin causing the feeling increasing similar to real gravity inside the spaceship.

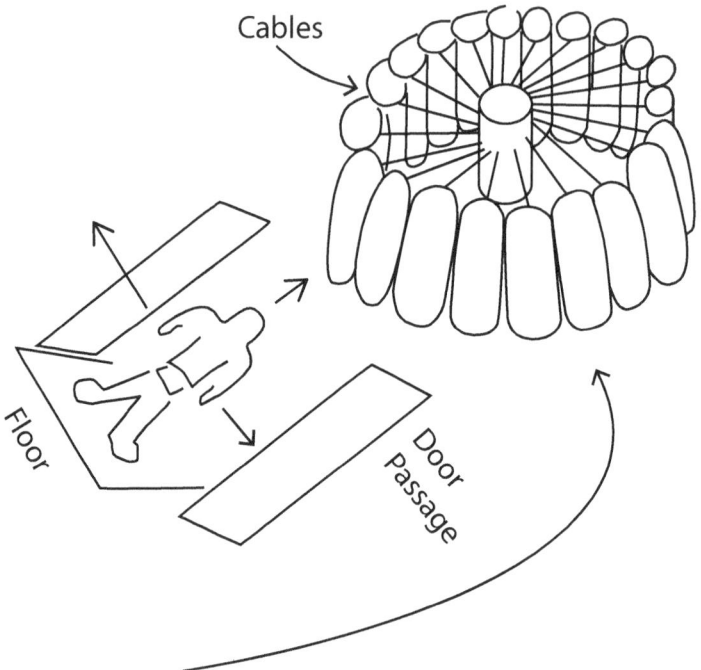

It's just a huge assembly
For the spin radius to be big

Cables

Floor

Door Passage

Hence modules lobbed into orbit connected in this arrangement results in an orbit to orbit interplanetary spaceship designed to door link so the occupants can walk erect from module to module.

Improved Scramjet Design (Hypersonic)

The NASA Mach 20 Scramjet worked 35 seconds because it's a dumb design. The design works better why? Because H2O Burn is 80% radiation. With intermix expansion occurs inside the Ram.

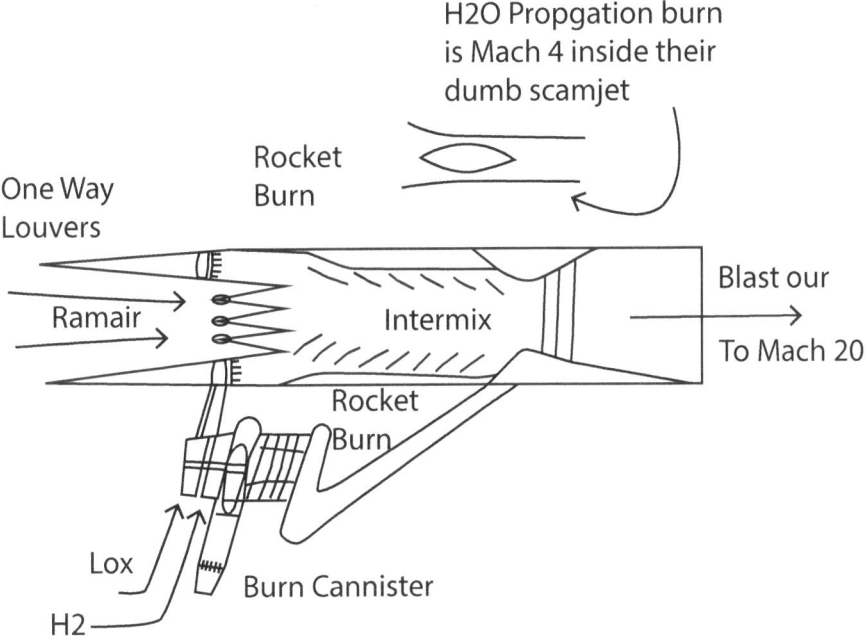

A fully retrievable booster rocket

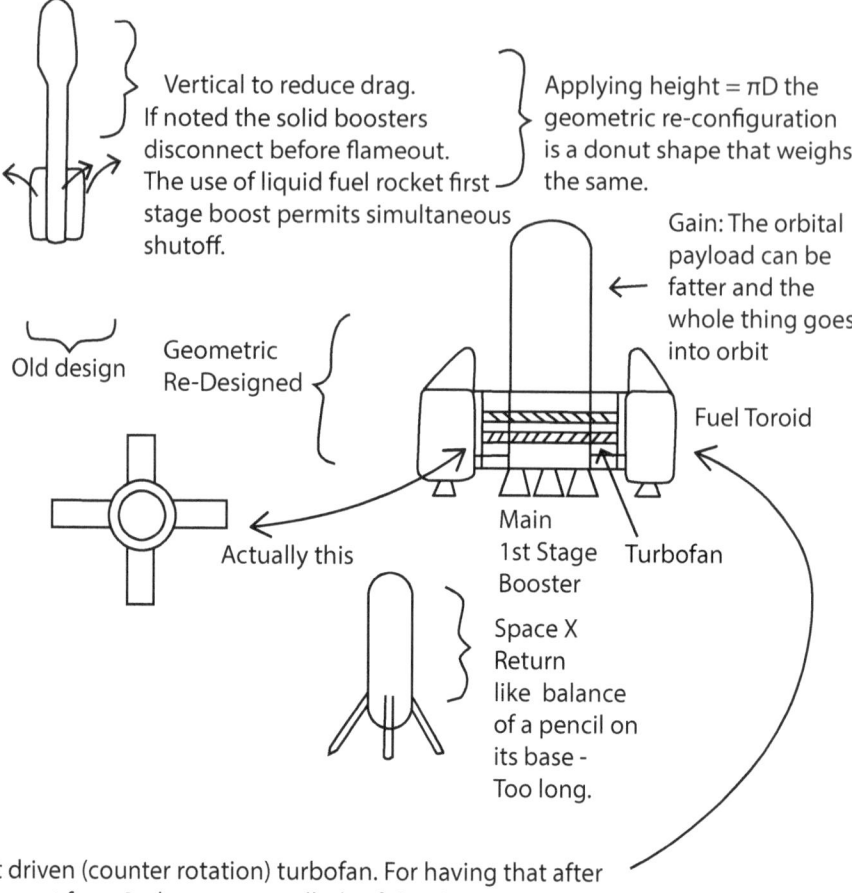

Vertical to reduce drag. If noted the solid boosters disconnect before flameout. The use of liquid fuel rocket first stage boost permits simultaneous shutoff.

Applying height = πD the geometric re-configuration is a donut shape that weighs the same.

Gain: The orbital payload can be fatter and the whole thing goes into orbit

Old design

Geometric Re-Designed

Fuel Toroid

Actually this

Main 1st Stage Booster

Turbofan

Space X Return like balance of a pencil on its base - Too long.

Rocket driven (counter rotation) turbofan. For having that after detachment from 2nd stage controlled soft landing (low center of gravity) Drone control return - with empty tanks for re-use.

More drag? Actually less because the counter roation return fan activated on liftoff contributes to lift in the amtosphere and shuts off to turn on again after stage separation.

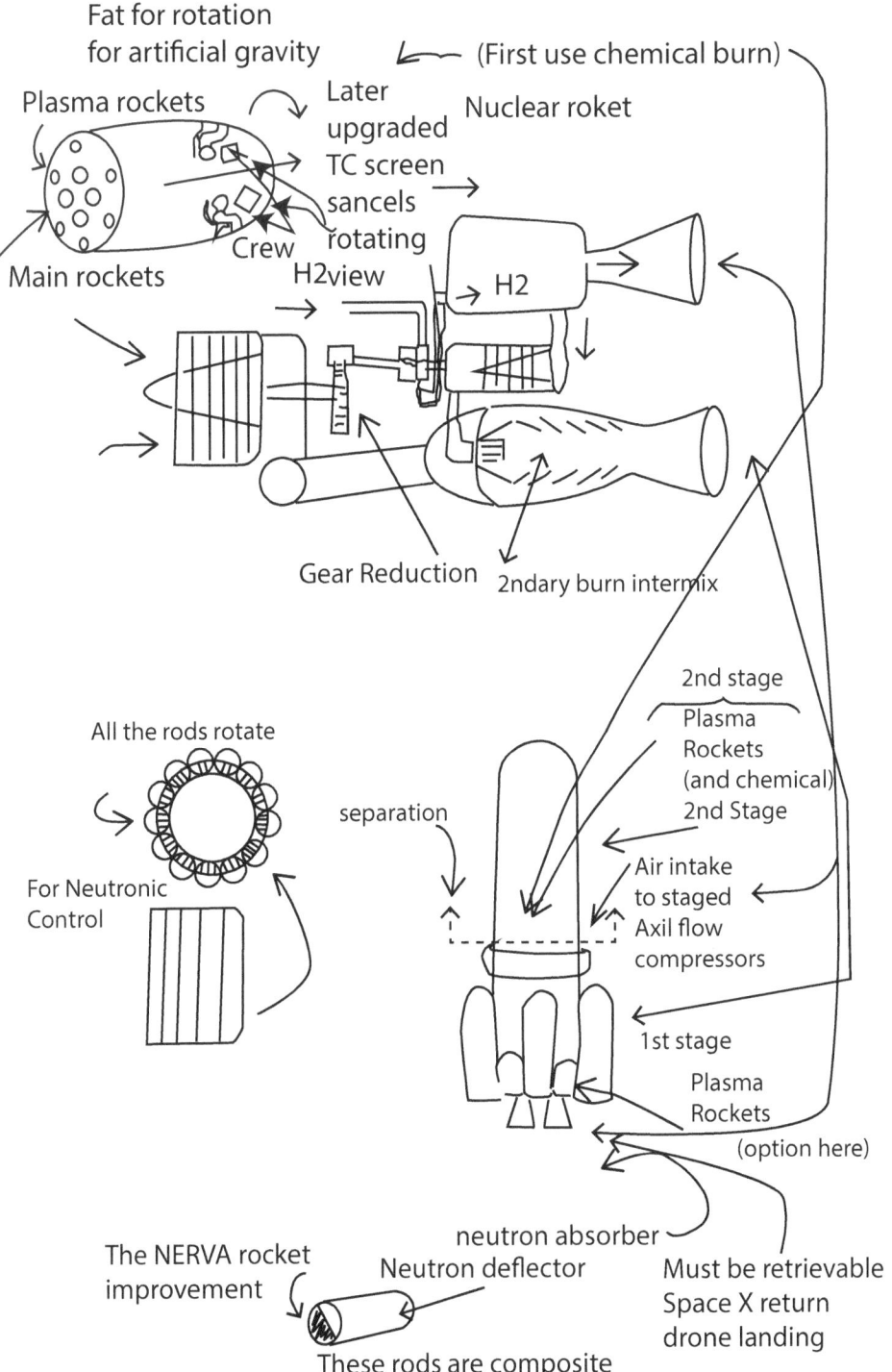

[This motor accounts for some UFO sightings]

It starts out as a pure jet - changes gradually with altitude to rocket (pure rocket) liquid.

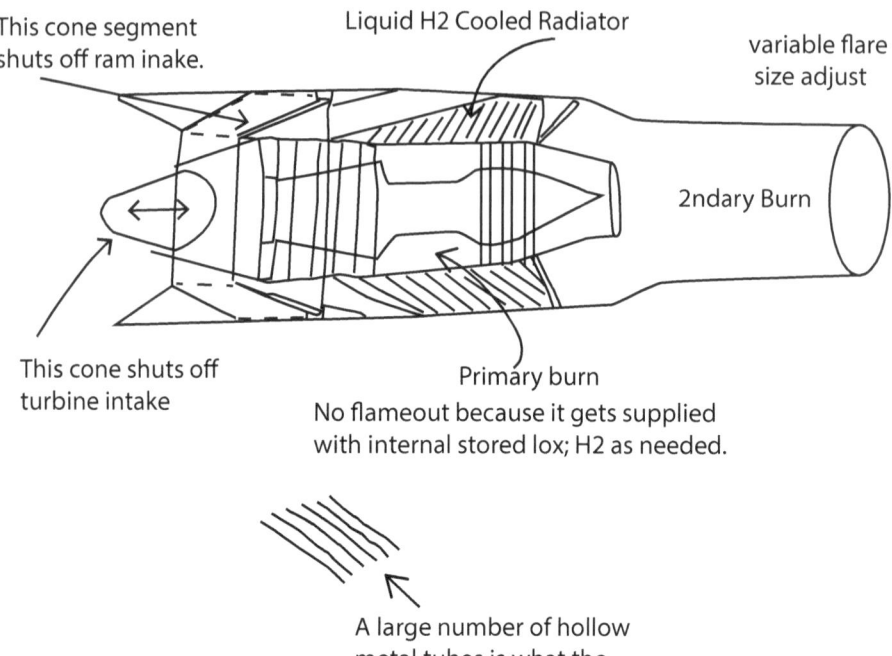

This cone segment shuts off ram inake.

Liquid H2 Cooled Radiator

variable flare size adjust

2ndary Burn

This cone shuts off turbine intake

Primary burn
No flameout because it gets supplied with internal stored lox; H2 as needed.

A large number of hollow metal tubes is what the radiator is made of.

Past 60 miles the jet is rocket because both cones move to block back gas from happening - By the a large percentage of orbital velocity attained so it takes little burn with intake closed to get orbital at that point.

It has developed in secret by both the USA and Russia since the 1980s if not earlier.

Americium Rocket

Internally the fluorescent lamp looking ring possibly was americium coated wich heated a propellant gas exiting through several (here one) rocket nozzle.

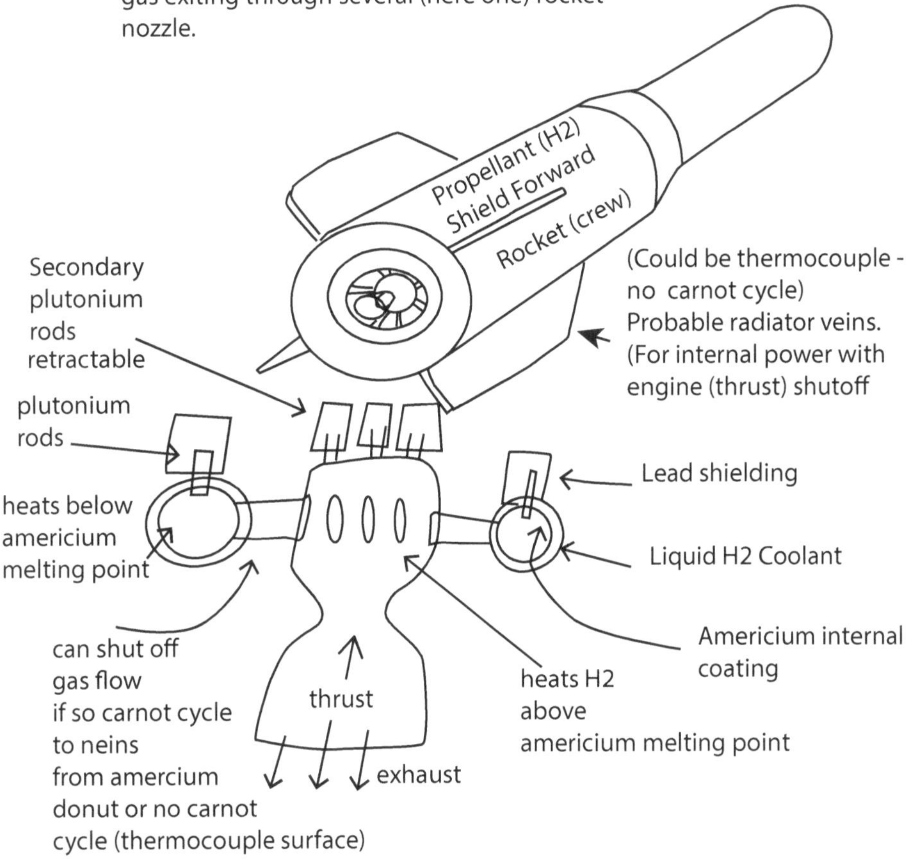

Propellant (H2)
Shield Forward
Rocket (crew)

Secondary plutonium rods retractable

plutonium rods

heats below americium melting point

can shut off gas flow if so carnot cycle to neins from amercium donut or no carnot cycle (thermocouple surface)

thrust

exhaust

(Could be thermocouple - no carnot cycle) Probable radiator veins. (For internal power with engine (thrust) shutoff

Lead shielding

Liquid H2 Coolant

Americium internal coating

heats H2 above americium melting point

Why americium? With plutonium rods retracted no radioactivity.

Interior gas circulation
rocket veins
electrical power out (instrumentation)
thermocouple surface

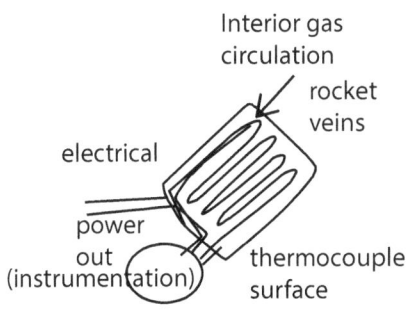

Counter-Rotating Fan Flying Top Vehicle

The vertical lift craft idea that Bell Helicopter rejected in 1973
The 4 blade dron design will work but I like this design better.

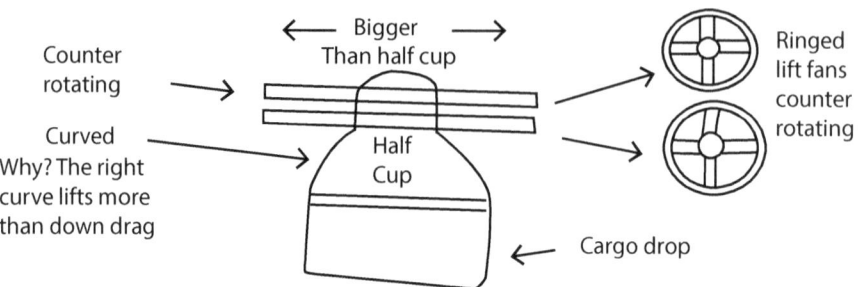

You see the induction motor at 60 cycles AC is 33% efficient (example house fan) but what Tesl did was raise frequancy. So then the device is hybrid - the gasoline or methane motor charges up lithium or sodium batteries why? No crashes. Motor shutoff and the device descends on battery power, varying blade pitch is how it moves forward. Ringed for blade edge visibilty. Done? Yes. Can it carry passengers? Yes. Will the idea be rejected? Yes. The rich pay nothing for ideas but make their money building and selling a product. Increased frequency increases power and reduces motor weight) - Telsa motors has those motors already built.

Gino Vesan

Flying vehicle commerce cuts down on highway traffic. Fewer raods needed. A lot of places not feasable to build roads.

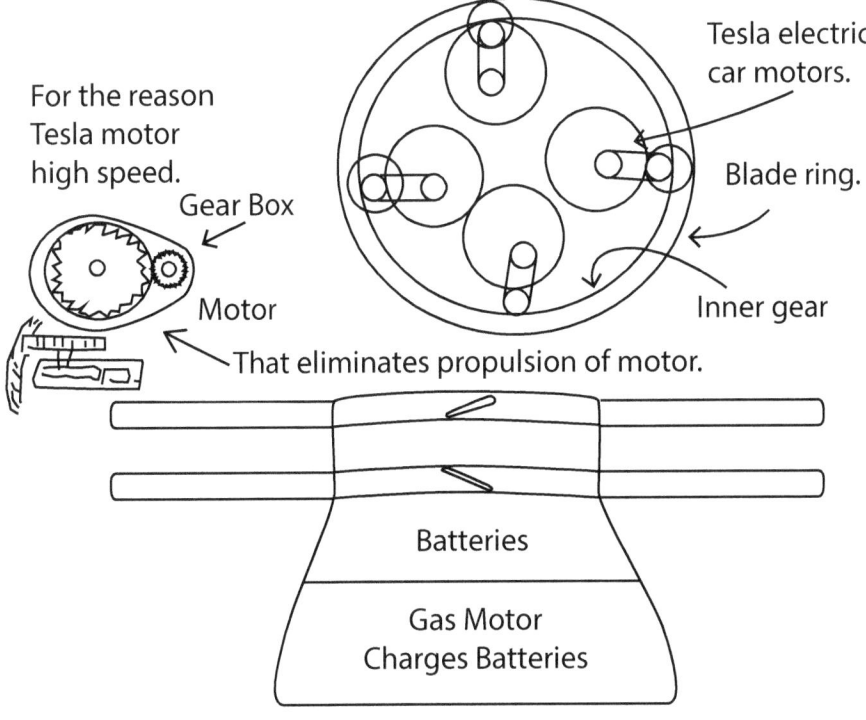

For the reason Tesla motor high speed.

Gear Box

Motor

Tesla electric car motors.

Blade ring.

Inner gear

That eliminates propulsion of motor.

Batteries

Gas Motor Charges Batteries

Batteries permit emergency control soft landing if gasoline motor quits.

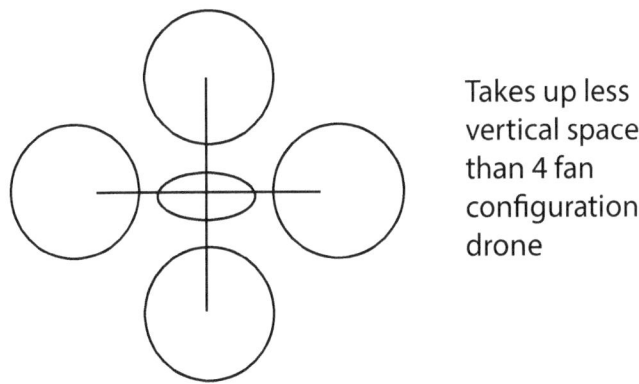

Takes up less vertical space than 4 fan configuration drone

It's the warbucks guy's who decide what gets built - us little guys don't get anything.

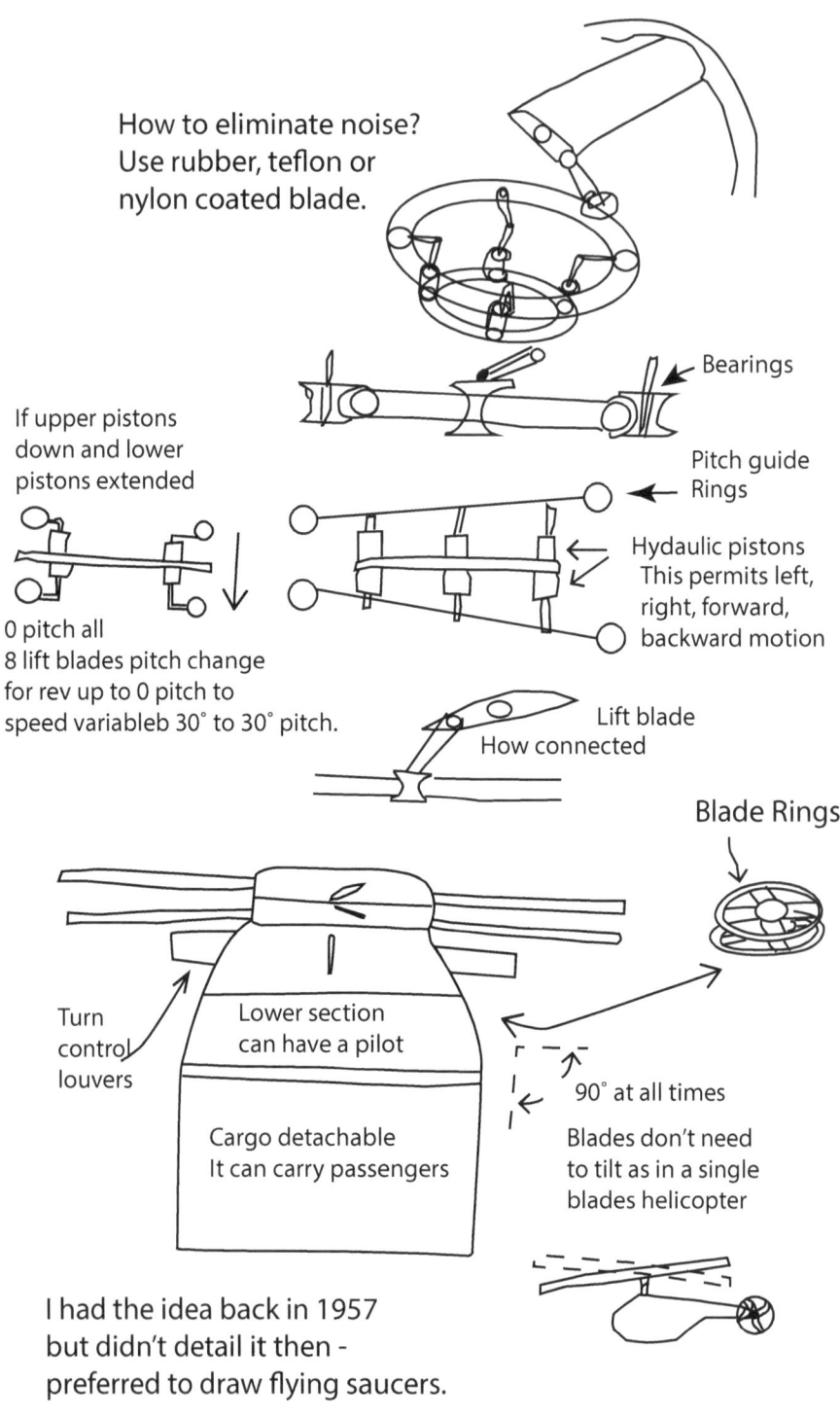

Slowly red shifts in 15 Billion Years
What started its travel as light

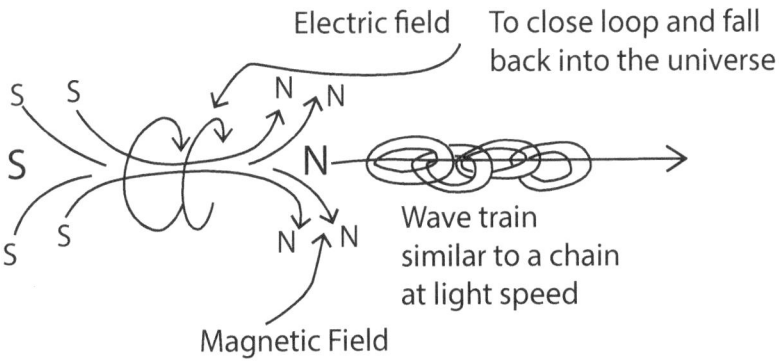

Electric field — To close loop and fall back into the universe

Wave train similar to a chain at light speed

Magnetic Field

Past 18,000 cycles/sec
The magnetic field doesn't have the sufficient to close loop why? because Einstein said At light speed light - the time is what?

The Steady State Theory has the big bang theory beat.

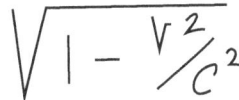

$$\sqrt{1 - v^2/c^2}$$

My clock works
Your clock Stopped
No My clock works Your clock doesn't work!
Whose clock stopped? <u>Neither</u>

Like a clock that moving relative to the observer.

Hubble didn't understand what red shift was.

So after 15 or so billion year - <u>closed loop!</u>

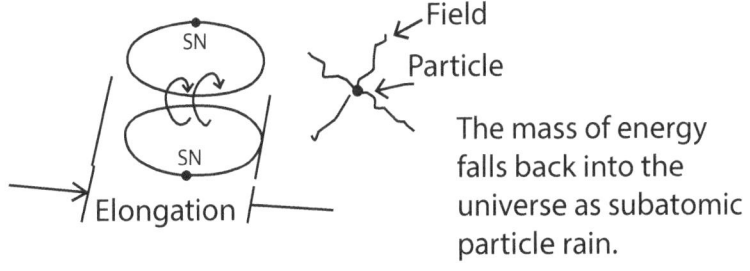

Field
Particle
Elongation

The mass of energy falls back into the universe as subatomic particle rain.

www.ingramcontent.com/pod-product-compliance
Lightning Source LLC
Chambersburg PA
CBHW040231220526
45473CB00001B/199